The Literary Structure
of Scientific Argument

D0146914

The Literary Structure of Scientific Argument

Historical Studies

Edited by Peter Dear

upp

University of Pennsylvania Press

Philadelphia

Library of Congress Cataloging-in-Publication Data
The Literary structure of scientific argument : historical studies /
    edited by Peter Dear.
        p.   cm.
    Includes bibliographical references and index.
    ISBN 0-8122-8185-3
    1. Science—History.   2. Science—Study and teaching—History.
3. Journalism—History.   4. Technical writing—History.   I. Dear,
Peter Robert.
Q126.8.L58   1991
509—dc20   90-45803
                                                      CIP

Contents

vi   Contents

Acknowledgments

THIS book emerged from a session on "The Literary Structure of Scientific Argument" that I organized at the 1987 meeting of the History of Science Society, held at Raleigh, North Carolina. I would like to thank Patricia Smith, Acquisitions Editor at the University of Pennsylvania Press, for her role in suggesting the creation of an edited collection along the same lines, and for her constant endeavors since—not least among them, regular promptings to come up with a better title (there have been about four). I would also like to thank all the contributors, who unfailingly worked to meet deadlines and fulfill requests, thus misleading me as to the normal rigors of editing. Tom Broman deserves special mention for countless conversations and ideas on these and related themes.

P.R.D.

Peter Dear

Introduction

JUST AS THE "LITERARY TURN" in intellectual history is by now well advanced,[1] so too is the extension of literary studies and theories into the domain of the history of science. Members of language and literature departments have begun to encompass the previously sacrosanct territory of the scientific paper within the broader confines of an understanding of the structure and functions of texts in general; some of their work is informed by structuralist and poststructuralist literary theory,[2] and much has developed from a renewed interest in rhetoric.[3] A book reviewer

1. See, e.g., Dominick LaCapra and Steven Kaplan, eds., *Modern European Intellectual History: Reappraisals and New Perspectives* (Ithaca, N.Y.: Cornell University Press, 1982), or the recent exchange in *American Historical Review* 94 (1989) between David Harlan, "Intellectual History and the Return of Literature," pp. 581–609, and David A. Hollinger, "The Return of the Prodigal: The Persistence of Historical Knowing," 610–621 (and Harlan, "Reply to David Hollinger," 622–626). One of the leading figures in these developments is Dominick LaCapra: see most recently his *Soundings in Critical Theory* (Ithaca, N.Y.: Cornell University Press, 1989). A concern with language is also appearing in social history: see Peter Burke and Roy Porter, eds., *The Social History of Language* (Cambridge: Cambridge University Press, 1987), esp. "Introduction" by Peter Burke, 1–20.

2. See, most recently, Stuart Peterfreund, ed., *Literature and Science: Theory and Practice* (Boston: Northeastern University Press, 1990); also a number of essays in Andrew E. Benjamin, Geoffrey N. Cantor, and John R. R. Christie, eds., *The Figural and the Literal: Problems of Language in the History of Science and Philosophy, 1630–1800* (Manchester: Manchester University Press, 1987), and the Foucauldian, structuralist study by Wilda C. Anderson, *Between the Library and the Laboratory: The Language of Chemistry in Eighteenth-Century France* (Baltimore and London: Johns Hopkins University Press, 1984). Classics such as Stanley Hyman's *The Tangled Bank: Darwin, Marx, Frazer and Freud* (New York: Atheneum, 1962), Marjorie Hope Nicholson's work on seventeenth- and eighteenth-century English science, or even (to a degree) recent work by G. S. Rousseau on the eighteenth century, fall more squarely into older traditions of literary studies.

3. For example, many of the contributions in John S. Nelson, Allan Megill, and Donald N. McCloskey, eds., *The Rhetoric of the Human Sciences: Language and Argument in Scholarship and Public Affairs* (Madison: University of Wisconsin Press, 1987), which, despite the title, contains essays on mathematics and natural science; Charles Bazerman, *Shaping Written Knowledge: The Genre and Activity of the Experimental Article in Science* (Madison: University of Wisconsin Press, 1988), with many additional references; "Symposium: Rhetoricians on the Rhetoric of Science" in *Science, Technology, & Human Values* 14 (1989): 3–49 (including an "Introduction" by Charles Bazerman, 3–6, and Steve Woolgar, "What is the Analysis of Scientific Rhetoric For? A Comment on the Possible Convergence Between Rhetorical Analysis and Social Studies of Science," 47–49, which points up the ambiguities of the rhetorical "project"). Some of this work, especially Bazerman's, draws on the functionalist sociology of science associated with the name of Robert K. Merton.

recently discerned, with apparent approval, the movement of "science and literature" towards the status of a "distinct academic discipline,"[4] and indeed there now exists a Society for Literature and Science devoted to the literary dimensions of science itself as well as to the intersection of science with literature more traditionally conceived. But the creation of specialties brings with it serious disadvantages, most notably a tendency to justify blinkered scholarly horizons.

Anglophone historians of science have not ignored the situation, of course; it is noteworthy, for example, that the review just mentioned appeared in the History of Science Society's journal, *Isis*. Nonetheless, the intellectual cachet that has attached itself in some quarters to literary theory serves to polarize opinion and practice in the history of science as in history more generally, and partly for this reason many historians of science have been slow in benefiting from the attention now being directed at the literary dimensions of scientific enterprises. Another source of reluctance, this time springing from a longstanding mainstream intellectualist approach to the subject, has also until recently attended the most penetrating examination of scientific textual practices now available to historians: that developed by sociologists of scientific knowledge. The classic study of contemporary laboratory scientific activity by Bruno Latour and Steve Woolgar, *Laboratory Life*,[5] devoted attention to the role of "literary inscriptions" in the practical creation of scientific knowledge and the way in which scientific papers, through particular textual strategies, process knowledge-claims by integrating them in stronger or weaker fashion with already accepted bodies of knowledge. These themes, taken up in more recent work, have pointed towards the significance of literary form and textual strategy in understanding the structure and behavior of scientific communities and the crucial role played by texts in the creation of knowledge.[6]

4. Paul Privateer, review of George S. Rousseau, ed., *Science and the Imagination* (New York: Annals of Scholarship, 1986), in *Isis* 80 (1989): 153–154; cf. G. S. Rousseau, review of Walter Schatzberg, Ronald A. Waite, and Jonathan K. Johnson, eds., *The Relations of Literature and Science: An Annotated Bibliography of Scholarship, 1880–1980* (New York: Modern Language Association, 1987), in *Isis* 80 (1989): 361–363.

5. Bruno Latour and Steve Woolgar, *Laboratory Life: The [Social] Construction of Scientific Facts* (Beverly Hills, Calif. and London: Sage, 1979; 2d ed., Princeton, N.J.: Princeton University Press, 1986).

6. For example, Bruno Latour in Chapter 1 of his *Science in Action: How to Follow Scientists and Engineers Through Society* (Cambridge, Mass.: Harvard University Press, 1987); essays in Michel Callon, John Law, and Arie Rip, eds., *Mapping the Dynamics of Science and Technology* (London: Macmillan, 1987). There is some intersection here with work by rhetoricians: see, for an outstanding example of the latter, Greg Myers, *Writing Biology: Texts in the Social Construction of Scientific Knowledge* (Madison: University of Wisconsin Press, 1990). More radical sociological departures include Steve Woolgar, "Discovery: Logic and Sequence in a

Such work in both literary studies and the sociology of science does, however, find some parallels in the mainstream activities of historians of science.[7] In 1976 Martin Rudwick published an article investigating the development in the nineteenth century of a "visual language for geology"; subsequent work by the same historian has analyzed the implicit social structuring of Darwin's notebooks and the appropriate ways of "reading" the various genres of Victorian geological writing, from field notebooks to published papers.[8] Early-modern chemistry has provided another focus of attention for linguistic, literary, and textual dimensions of science, especially since the seminal work of Owen Hannaway, while Geoffrey Cantor has written on the nature of the metaphors infusing eighteenth-century British physical optics.[9] Frederic L. Holmes has pointed to the ways in

Scientific Text," in Karin D. Knorr, Roger Krohn, and Richard Whitley, eds., *The Social Process of Scientific Investigation* (Dordrecht: D. Reidel, 1981), 239–268; Michael Mulkay and Nigel Gilbert, *Opening Pandora's Box: A Sociological Analysis of Scientists' Discourse* (Cambridge: Cambridge University Press, 1984); Mulkay, *The Word and the World: Explorations in the Form of Sociological Analysis* (London: George Allen and Unwin, 1985); essays in Steve Woolgar, ed., *Knowledge and Reflexivity: New Frontiers in the Sociology of Knowledge* (Beverly Hills, Calif. and London: Sage, 1988).

7. Jan Golinski, "Language, Discourse and Science," in R. C. Olby, G. N. Cantor, J. R. R. Christie, and M. J. S. Hodge, eds., *Companion to the History of Modern Science* (London and New York: Routledge, 1990), 110–123, reviews some of the relevant issues and a selection of literature. Golinski rightly observes, however, the reluctance of historians of science to make this move. Even the recent collection mentioned above, *The Figural and the Literal*, is primarily concerned with language *per se* and with literary theory; it focuses only peripherally on the dynamics of scientific communities or on the creation of the actual content of scientific knowledge. Thomas Kuhn's observations on the function of textbooks in his *The Structure of Scientific Revolutions*, 2d ed. (Chicago: University of Chicago Press, 1970), are, of course, well known, but the exploitation of the idea by historians of science has been quite restricted. In some quarters Michel Foucault's notion of "discourse" has had a degree of impact (especially *The Order of Things: An Archaeology of the Human Sciences* [New York: Pantheon Books, 1971] and "The Discourse on Language," in *The Archaeology of Knowledge and the Discourse on Language*, trans. A. M. Sheridan Smith [New York: Pantheon Books, 1972]). Its perceived subversive character vis-à-vis the historian's craft, however, has meant that its direct usefulness has been only dimly felt—which is not to deny its role in sensitizing some historians to the importance of language.

8. Martin J. S. Rudwick, "The Emergence of a Visual Language for Geological Science, 1760–1840," *History of Science* 14 (1976): 149–195; idem, "Charles Darwin in London: The Integration of Public and Private Science," *Isis* 73 (1982): 186–206; idem, *The Great Devonian Controversy: The Shaping of Scientific Knowledge Among Gentlemanly Specialists* (Chicago: University of Chicago Press, 1985), esp. 429–435.

9. Owen Hannaway, *The Chemists and the Word: The Didactic Origins of Chemistry* (Baltimore and London: Johns Hopkins University Press, 1975); J. R. R. Christie and J. V. Golinski, "The Spreading of the Word: New Directions in the Historiography of Chemistry, 1600–1800," *History of Science* 20 (1982): 235–266; Jan V. Golinski, "Peter Shaw: Chemistry and Communication in Augustan England," *Ambix* 30 (1983): 19–29; idem, "Robert Boyle: Scepticism and Authority in Seventeenth-Century Chemical Discourse," in Benjamin et al., *The Figural and the Literal*, 58–82; cf. also Anderson, *Between the Library and the Laboratory*. The potential fruitfulness of chemistry in this regard was always evident in the explicit nature of the issues in contemporary, particularly eighteenth-century, chemical texts: cf. Maurice Crosland,

which scientific arguments are constructed through the casting and recasting of papers that combine and emphasize material and ideas in new ways that did not spring full-blown from the laboratory work on which they draw; looking particularly at successive drafts of some of Lavoisier's scientific memoirs, he stresses the role of writing in the thinking processes that lead to scientific discovery.[10] In addition, Steven Shapin, exploiting the sociological work mentioned above, has applied its insights to properly historical questions in making sense of Robert Boyle's writings through consideration of the nature of the philosophical community that Boyle wanted to establish.[11]

These kinds of studies indicate a growing awareness by historians of science of the potentialities of attention to language, rhetoric, and textual forms in understanding how science has been created. They signal an increasing sensitivity to the lesson, developed in other disciplines, that language is not simply a transparent medium of communication, but a shaper (perhaps a realizer) of thought and an embodiment of social rela-

---

Historical Studies in the Language of Chemistry (Cambridge, Mass.: Harvard University Press, 1962; New York: Dover, 1978). Geoffrey Cantor, "Light and Enlightenment: An Exploration of Mid-Eighteenth-Century Modes of Discourse," in David C. Lindberg and Geoffrey Cantor, *The Discourse of Light from the Middle Ages to the Enlightenment* (Los Angeles: Clark Memorial Library, 1985), 67–106; idem, "Weighing Light: The Role of Metaphor in Eighteenth-Century Optical Discourse," in Benjamin, *The Figural and the Literal*, 124–146.

10. Frederic L. Holmes, *Lavoisier and the Chemistry of Life: An Exploration of Scientific Creativity* (Madison: University of Wisconsin Press, 1985); idem, "Scientific Writing and Scientific Discovery," *Isis* 78 (1987): 220–235.

11. Steven Shapin, "Pump and Circumstance: Robert Boyle's Literary Technology," *Social Studies of Science* 14 (1984): 481–519, and Shapin and Simon Schaffer, *Leviathan and the Air-Pump: Hobbes, Boyle and the Experimental Life* (Princeton, N.J.: Princeton University Press, 1985); also for related concerns, Peter Dear, "*Totius in verba*: Rhetoric and Authority in the Early Royal Society," *Isis* 76 (1985): 145–161. The most important recent addition to the continually growing body of literature on the early Royal Society and "prose style," although not traditionally a central part of the concerns of historians of science, is Brian Vickers, "The Royal Society and English Prose Style: A Reassessment," in Vickers and Nancy S. Struever, *Rhetoric and the Pursuit of Truth: Language Change in the Seventeenth and Eighteenth Centuries* (Los Angeles: Clark Memorial Library, 1985), 3–76.

A very suggestive and original article on the role of genre in the history of science has received surprisingly little attention: Edward Grant, "Aristotelianism and the Longevity of the Medieval World View," *History of Science* 16 (1978): 93–106, on the implications of the commentary genre in natural philosophy. Also noteworthy is Loren Graham, Wolf Lepenies, and Peter Weingart, eds., *Functions and Uses of Disciplinary Histories* (Dordrecht: D. Reidel, 1983). This collection, however, concentrates almost exclusively on the twentieth century and has little to say about literary form, genre considerations, or textual issues, but focuses instead on the content of disciplinary histories and their functions in establishing or legitimating particular scientific specialties or disciplinary approaches and the communities that embody them. The disposition of that content, and the literary parameters that help to determine it, are not addressed. The essays in the present collection provide references to additional relevant literature.

tions. That lesson remains less than a commonplace, however, no doubt because of the well-entrenched, and functional, place of the opposite assumption in scientific practice itself, where words refer unproblematically and scientific rhetoric reproduces uncluttered inferential reasoning.[12] The essays in the present collection attempt to demonstrate by example the territories of genuine historical insight (rather than just artful redescription) awaiting the properly attuned explorer, as well as present specific findings of value for a broad sweep of issues in the history of science.

By covering a wide variety of scientific disciplines, including medicine and mathematics as well as physiology and physics, and by spanning three centuries, these studies not only address problems proper to each, but also display important aspects of commonality. An incomplete summary of their thematic repertoire might look like this:

1) The role of genres in perpetuating, changing, or subverting scientific research programs.
2) The role of genres in defining disciplinary boundaries.
3) The role of scientific texts in embodying the cognitive assumptions or social structure of the sciences to which they belong.
4) The ways in which literary forms can direct the cognitive content of a science through constraining problem-choice or through requiring (via their own disciplinary entrenchment) particular kinds of theoretical and experimental formulation.

These four related issues recognize the fundamentally interactive (and indeterminate) nature of the relationship between textual and other practices in scientific activity.

Despite the temporal scope of this book, the essays are not ordered chronologically. Such a presentation would have obscured the larger purposes of the volume, which are best brought out by a grouping that renders more evident the issues just enumerated. Part I, therefore, contains discussions of issues related above all to genre and disciplinary structure, implicating especially points 1) and 2). Part II, somewhat in the manner of a transition, comprises an examination of a textually delineated shift in the epistemological posture of a discipline, together with the concomitant reorientations of its social workings and traditions of practice, bringing in points 1), 3), and 4). Part III consists of essays focused especially on textual form and strategy in the act of creating specific pieces of scientific knowl-

12. Reflective self-awareness by scientists certainly exists, of course: see Roald Hoffmann, "Die chemische Veröffentlichung—Entwicklung oder Erstarrung im Rituellen?" *Angewandte Chemie* 100 (1988): 1653–1663.

edge, and related most closely to points 3) and 4). The parts cannot, however, be regarded as thematically distinct, insofar as the first also illuminates point 3), relating to cognitive assumptions and social structure, and the third frequently invokes genre considerations in addressing the meaning of textual forms. Nonetheless, the groupings assist in emphasizing the larger lessons of each essay through juxtaposition with those others most closely related to it.

Broman's piece fittingly opens Part I with a discussion of genre and its handling in literary studies. The ideas there laid out establish clearly some of the parameters not only for Broman's investigation of Reil's physiology journal and its subsequent mutations, but also for Nyhart's essay on a late nineteenth-century genre of zoology and its changing social/disciplinary meaning, and for Hunt's analysis of the rhetorical hallmark of "rigor" embossed on British mathematical writing around the turn of the century. Broman argues that genres of scientific writing are more than either conventional packaging for extratextual reasoning or reflections of the social practices of a scientific community. They also actively shape and constrain the kind of knowledge that they purvey—different genres, he claims, will support different kinds of theories. Changes in dominant genres of scientific writing can thus render redundant older theoretical constructs and promote the appearance of newer ones. At the same time, in the specific case of Reil's *Archiv*, Broman shows the response of the journal's structure to broader disciplinary and institutional changes, particularly concerning the relation of physiology to medical education. The move from textbook to journal paralleled a subtle shift in the character of another genre, that of the medical dissertation, and in a sense reified it.

Nyhart's spotlighting of a case from zoology later in the century reveals the play of other forces. Editorial initiatives and policies are just one element in the career and meaning of a journal. Nyhart traces the development of the *Zeitschrift für wissenschaftliche Zoologie* from a journal explicitly conceived as a sort of cooperative disciplinary workshop, through the unintended disruptive role of priority disputes enabled by precisely that journal structure, to the irony of the institutionalization of published dispute as an appropriate rhetorical form of self-establishment in a new kind of disciplinary community: a community the commonality of which lay in deployment of *topoi* proclaiming recognized norms as badges of allegiance. Finally, Hunt shows how to understand a particular episode, the irregular rejection by the Royal Society of a paper sent for publication by Oliver Heaviside, by constructing around that event the meanings attaching to Heaviside's forms of argument and those of the British mathematical community at about the turn of the century. The genre-characteristic of "rigor" was of paramount importance to pure mathematicians because it set them apart from the physical scientists who had until recently, especially through the Mathematical Tripos at Cambridge, held mathematics in thrall to physics. The work presented by Heaviside addressed areas of mathematics held in disrepute by the newly autonomous mathematicians, and their repudiation was rooted in its resistance to being cast in the appropriate genre of rigorous deductive argument. Heaviside, the freelance, saw no reason to ignore interesting and potentially valuable results merely because they could not, at least yet, be presented in a particular form. The professionals, however, ensured that he was denied publication in a forum that they wished to control—and that helped them to define their discipline.

Lissa Roberts's contribution in Part II provides a sensitive reading of eighteenth-century chemical tables to show how superficial similarities in fact mask changes that go to the heart of the "Chemical Revolution." Her story involves consideration of the meaning of making chemistry a science rather than an art (in the terms of the period). The solution to the problem of accomplishing that redefinition, so as to allow chemistry a disciplinary status desirable to its practitioners, set up its own difficulties and precipitated the new solution of Lavoisier and his allies. This process, Roberts argues, may be read in the tables used to capture the products of the discipline. Those tables constituted both what counted as chemical knowledge and how such knowledge was properly to be made; furthermore, they defined what chemistry was *about*—the "nature" that was its object. The relation between technical art and nonartifactual nature that lay at the root of the original art/science opposition initiating Roberts's account is shown, through a reading of Lavoisier's tables, to return to a new point of dialectical resolution whereby chemistry has become a new kind of enterprise.

Part III assembles studies that focus on the construction of specific pieces of scientific knowledge. All three examine in detail the textual strategies through which knowledge appeared and by which it was constituted *as* knowledge, attempting to afford an understanding of the literary structure of different scientific practices in different times and places by reference either to their epistemological or to their institutional characteristics. In addition, all three concentrate specifically on experiment as the site of knowledge creation.[13] Addressing the very question of what constitutes an

13. Experimental practice in science has recently come to the fore in studies of its history, philosophy, and sociology. Among recent works of special historical relevance, see Shapin and

"experiment," Dear examines the nature of some apparently "experimental" science in the seventeenth century to expose the irredeemably linguistic dimension of that conception. By discussing both routine practice and the rhetorical resources of controversy among Galileo and Jesuit scholastic writers concerning phenomena of free fall, the essay shows how the intersection of epistemological discourse, typically validated by references to Aristotle, with argument aimed at establishing specific knowledge claims, gave to narratives about the performance of contrived physical events meanings and rhetorical functions very different from those of later forms of experimentalism that denied the expectations of Aristotelianism. Holmes, in turn, looks at work in chemistry performed by members of the Royal Academy of Sciences in Paris during the decades around 1700. Holmes finds that the work done in the Academy developed techniques and conventions of combining, or differentially emphasizing, narrative and formalized argument in ways that came to typify the modern experimental research article—ways that differ markedly from those of the early Royal Society of London, which deployed dense narrative as their chief textual characteristic. Accounting for the formal properties of the Academy's practices, Holmes points to the social and institutional features of their setting, where members wrote for fellow specialists as part of prolonged series of full-time, effectively professional, experimental research. The exigencies of these conditions of textual production helped to displace the narrative structure that Holmes argues is intrinsic to experimental papers from a simple "investigative narrative" to the conventionalized narrative codified in modern guides to the genre of the research report.

Lisa Rosner's concluding essay on models of experiment in the medical world of eighteenth-century Edinburgh shows how "experimental teaching, like experiment itself, was shaped by the available models for presentation of experiential knowledge." This was a reciprocal interaction: the chief models from which students (and hence practitioners) learned to understand the meaning of experiment—its textual and rhetorical place—were the lecture-demonstration, a kind of "live performance" that placed the experimental event in a rhetorical setting mirrored in the medical dissertation, and medical case histories, from which the student was expected to be able to diagnose on the basis of a textual account rather than an actual

---

Schaffer, *Leviathan;* Diderick Batens and Jean Paul Van Bendegem, eds., *Theory and Experiment: Recent Insights and New Perspectives* (Dordrecht: D. Reidel, 1988); and, above all, David Gooding, Trevor Pinch, and Simon Schaffer, eds., *The Uses of Experiment: Studies in the Natural Sciences* (Cambridge: Cambridge University Press, 1989), with extensive bibliography.

examination. These also reflected the more generally pervasive precedent of experimental narrative and its availability for argument that stemmed from seventeenth-century British models.

Rosner's piece, like the others in this collection, can be seen as a kind of *cultural* history of science.[14] The work of these authors, with their refusal to treat the scientific text as a mere informational commodity, on the one hand, or a transparent medium of thought, on the other, gives the lie to the still prevalent view of the history of science as divided into the polar opposites of "social" and "intellectual." Because intellectual life, like all activities, is necessarily (and in that sense trivially) social, it is impossible to cordon off from its history the various means by which individuals interact so as to create meaning and knowledge. The contributors draw variously on lessons from literary theory and the sociology of science to confront their own properly historical concerns; what they study are the cultures of scientific communities as read in their texts.

14. The best examples of history of science presented explicitly as a part of cultural history tend to concern Victorian Britain: see, for example, James A. Secord, *Controversy in Victorian Geology: The Cambrian-Silurian Dispute* (Princeton, N.J.: Princeton University Press, 1986). This is only fitting, given the subject matter of Susan Faye Cannon's call-to-arms, *Science in Culture: The Early Victorian Period* (New York: Science History Publications, 1978). Cultural history itself, led especially by Roger Chartier, is now focusing on texts as objects in their own right: see, for example, Roger Chartier, *Cultural History: Between Practices and Representations* (Ithaca, N.Y.: Cornell University Press, 1988); Lynn Hunt, ed., *The New Cultural History* (Berkeley: University of California Press, 1989).

Part I

Genre and Discipline

Thomas H. Broman

1. J. C. Reil and the "Journalization" of Physiology

THE CONFLICT between "internal" and "external" history of science appears to have subsided, fortunately, but the problems raised by it remain as troublesome as ever. Although historians have fruitfully explored topics such as the rhetorical adaptation of science to its audiences, the dynamics of scientific investigations, and the proliferation of scientific disciplines and institutions, other topics in the history of science have proven themselves less tractable. Especially when they have turned to studying scientific ideas— the "cognitive content" of science, as it is known in some circles—historians have found it difficult to show how given ideas are rooted in particular historical circumstances. This point was underscored recently by Charles Rosenberg, who drew upon his experience as editor of *Isis* to claim that most articles demonstrate "how few historians of science succeed in relating ideas to social and institutional contexts."[1]

In this essay, I want to address this problem at two levels. First, I will argue that studying the genres of scientific writing provides a device for situating ideas in their historical context. The utility of genres for this purpose lies in the fact that genres both serve identifiable social functions and act to structure the material they present in characteristic ways.[2] These

---

1. Charles Rosenberg, "Woods or Trees? Ideas and Actors in the History of Science," *Isis* 79 (1988): 565.
2. It should be noted that historians of medieval and early modern science, by virtue of the pronounced exegetical orientation of scholarship in those periods, have long attended to the generic properties of the scholarly texts. Owen Hannaway pioneered this kind of interpretation in his now-classic *The Chemists and the Word: The Didactic Origins of Chemistry* (Baltimore: Johns Hopkins University Press, 1975). See also Edward Grant, "Aristotelianism and the Longevity of the Medieval World View," *History of Science* 16 (1978): 93–106; Wilda C. Anderson, *Between the Library and the Laboratory: The Language of Chemistry in Eighteenth-Century France* (Baltimore: Johns Hopkins University Press, 1984); Andrew Wear, "William Harvey and the 'Way of the Anatomists'." *History of Science* 21 (1983): 223–249; Andrew Cunningham, "Fabricius and the 'Aristotle Project' in Anatomical Teaching and Research at Padua," in Andrew Wear, et al., eds., *The Medical Renaissance of the Sixteenth Century* (Cambridge: Cambridge University Press, 1985), 195–222; Jerome Bylebyl, "Disputation and Description in the Renaissance Pulse Controversy," in *The Medical Renaissance of the Sixteenth*

twin properties allow genres to link what is produced in the mind with the world in which those products find their place. Second, to illustrate how this type of analysis might be applied to the history of science, I will describe how the appearance of a new periodical, the *Archiv für die Physiologie,* both represented and furthered a decisive turn in the science of physiology at the end of the eighteenth century. As will become abundantly clear in the discussion, the interplay between intellectual innovation, sociocultural context, and literary genre is an exceedingly complex one, for which this essay can provide only a preliminary survey.

Genres in Intellectual History

Although the idea of grouping literature into types or genres is an ancient one, it has not always been the object of veneration. In fact, the utility of genre concepts in literary criticism has been vigorously debated for at least a century, and in describing genres it will be worthwhile to pause briefly with some of the arguments that have surrounded them. The shortcomings of generic treatments of literature have been assessed by widely differing authors, ranging from Benedetto Croce, who thought interpretation of literary works as members of generic groups did violence to the internal coherence and uniqueness of each piece of literature, to Roland Barthes, who argued that individual texts were indeterminate aggregations of conflicting discourses and generic codes.[3] Whatever their theoretical stance, criticisms of genre usually make the same basic argument: all taxonomies of genre depend on enumeration of formal characteristics to distinguish one

---

Century, 223–245; and Nancy Siraisi, *Avicenna in Renaissance Italy* (Princeton, N.J.: Princeton University Press, 1987). These works are exemplary in their use of textual characteristics in the interpretation of ideas, but they do not explicitly use genres as mediating agents between ideas and environment in the way I do here.

Recently, Charles Bazerman applied generic considerations to the experimental research report in *Shaping Written Knowledge: The Genre and Activity of the Experimental Article in Science* (Madison: University of Wisconsin Press, 1988). Bazerman's work is important for introducing historians to this approach, but it has several shortcomings. Most seriously, it fails to explain the development of its genre in a specific historical setting. Detailed criticism of Bazerman's work is contained in Peter Dear, "Sociology? History? Historical Sociology? A Response to Bazerman," *Social Epistemology* 2 (1988): 275–278. Finally, Ludmilla Jordanova directs some attention to genres of scientific writing in L. J. Jordanova, ed., *Languages of Nature: Critical Essays on Science and Literature* (London: Free Association Books, 1986), 22–26.

3. Heather Dubrow, *Genre* (London and New York: Methuen, 1982), 83–84. Dubrow's book is a useful introduction to the topic of genres and is especially helpful in its discussion of genre theory in the twentieth century.

kind from another, but since the characteristics are those of the genre and not necessarily of any individual text, one immediately encounters problems with the relationship between the genre and the texts that supposedly constitute it. No individual text ever contains all the marks of its genre; indeed, many texts mix different genres. Therefore, these classifications become ensnarled in unpleasant ontological questions of how texts "belong" to their genre(s).[4]

These criticisms arise, according to Ralph Cohen, because scholars have mistakenly attempted to define genres in terms of their formal properties. Cohen rejects this essentialism in favor of what he calls an "empirical" approach.[5] He argues that "genre concepts . . . arise, change, and decline for historical reasons," and consequently the grouping that produces a genre "is a process, not a determinate category."[6] Cohen's conception of genres as being situated in history is widely shared by other advocates of genre analysis who usually describe them as sets of conventions that make communication between writers and readers possible.[7] Fredric Jameson, for example, has written about genre as a "contract" between writers and their intended public that regulates various aspects of the literary product.[8] But whether they are called conventions, contracts, or "horizons of expectation,"[9] it is clear this way of looking at literary genres induces critics to examine them in concrete historical settings.

If only it were so simple. While in principle all this may sound straightforward, when one gets down to cases problems often arise. After all, what is the social "function" of cultural artifacts that exist in part for aesthetic enjoyment? What historical functions were served by the sonnet in Petrarch's day and in Shakespeare's own different time, and what historical conditions can be cited to explain its revival in the nineteenth century? The

4. Earl Miner, "Some Issues of Literary 'Species, or Distinct Kind'," in Barbara K. Lewalski, ed., *Renaissance Genres: Essays on Theory, History, and Interpretation* (Cambridge, Mass. and London: Harvard University Press, 1986), 15–44; Ralph Cohen, "History and Genre," *New Literary History* 17 (1986): 203–219. Given their disposition toward literature, it goes without saying that structuralists such as Northrop Frye have been among the most ardent defenders and classifiers of literary genres. See Frye's *Anatomy of Criticism* (Princeton, N.J.: Princeton University Press, 1957), 33–67, 243–251.

5. Cohen, "History and Genre," 210.

6. Ibid., 204.

7. Alastair Fowler, *Kinds of Literature: An Introduction to the Theory of Genres and Modes* (Oxford: Clarendon Press, 1982).

8. The idea of genre as contract is discussed in Fredric Jameson, "Magical Narratives: On the Dialectical Use of Genre Criticism," in Jameson, *The Political Unconscious* (Ithaca, N.Y.: Cornell University Press, 1981), 103–150.

9. Hans Robert Jauss, "Literary History as a Challenge to Literary Theory," *New Literary History* 2 (1970): 7–37.

problem may not be insurmountable, but persuasive explanations that are grounded in historical circumstance do not readily spring to mind.

The historian interested in working with genres might get better guidance from rhetoricians, the second group of scholars who make genres their concern. Although rhetoricians too have often scratched their heads over questions of formal classification, several writers have recently begun moving away from classification as the primary aim of genre studies, preferring to see rhetorical genres (such as the courtroom summation or the funeral oration) in pragmatic terms as responses to recurrent social situations.[10] This is a useful formulation, because it points directly to the functions served by genres, and it indicates how they can be situated historically. From the historian's perspective, the great advantage held by rhetoricians in this business is that they deal with *nonfictional* literature, and the social functions of such writings are far more accessible than those of fictional literature. Thus when we speak of rhetorical genres as establishing a bridge of communication between writer and audience, the circumstances that guide the construction of that bridge can be conceived with standard historical tools.

But although examining the function of genres in concrete historical settings is an important key to placing written knowledge in its context, it does not finish the job. For it is not enough to say what a genre *does;* we must also say what it *is.* In this way, Cohen's recommendation that we see genres as processes in historical context seems inadequate. Without paying attention to the problem of form, we would be making in effect an idealist argument wherein genres provide a means by which independently generated ideas acquire social meaning. Or, to return to the bridge metaphor, we would be describing the traffic that crosses the bridge—where it originates and finds its terminus, and what comprises it—without describing the bridge itself. Obviously, however, the bridge makes a difference, in part because it can help explain the type of traffic that crosses it. To take a less metaphorical example, to describe only the pedagogical function of a textbook would be to place it in the same category as the periodic table of the elements, or a ball-and-stick model of molecules. But a textbook is what it is

10. Carolyn R. Miller, "Genre as Social Action," *Quarterly Journal of Speech* 70 (1984): 151–167. The meaning of "recurrent situation" has been debated by several writers who have argued over whether such situations are objectively given or actively defined as such by participants. Miller inclines toward the latter, although she fails to see such definitions outside individual choice. Clearly, however, "definitions" could also be produced and reinforced by existing social, economic, and political institutions. On the nature of recurrent situations, see Lloyd F. Bitzer, "The Rhetorical Situation," *Philosophy and Rhetoric* 1 (1968): 1–14, and Kathleen M. Jamieson, "Generic Constraints and the Rhetorical Situation," *Philosophy and Rhetoric* 6 (1973): 162–170.

largely because of its structural qualities: what it contains and how it presents its contents. Any writer who sets out to write a textbook, or a dissertation or scientific research article for that matter, must take into account those conventions and practices that permit a writing to be recognized and accepted by readers as an exemplar of a particular genre. In studying genres, therefore, we must keep both perspectives—the functional and the formal or structural—firmly in mind.

Finally, we must consider the relationship between the forms of communication provided by genres and the ideas they contain. I previously said that genres "structure" the ideas contained in them, but how far does this structuring activity go? If scholarly genres embody institutionalized rules for doing scholarship (as I indicated above), then by writing in one genre as opposed to another to what extent is a writer expressing a different idea? This problem brings us to the heart of the matter, for if there is not at least a reasonably tight correspondence between the formal structures of written genres and their intellectual contents, the utility of genres for linking the internal and external elements dissolves. As I hope to make clear in my discussion of the *Archiv für die Physiologie*, genres of writing and scientific theories develop together and become established in particular historical circumstances. Once established, these pairings can be remarkably durable, so long as the genres continue to serve the same social function. There are historical moments, however, when changing historical circumstances cause the pairing to become loosened, and this creates a space in which new genre/theory pairings can develop.

In talking about genre/theory pairings, I should emphasize that I do not mean pairings between a scholarly genre and any *single* scientific theory. Rather, I am talking about particular *groups* of scientific theories that become paired with genres. It may be objected that I am in danger of falling into a tautology here: after all, what defines these theories as a "group" is their sharing of the structural elements (such as the rhetorical formats and linguistic patterns) of their genre. What does it gain us to say the genre defines what is in the genre? Yet, using the structural elements of the genre to group theories does not eliminate the way in which the concepts "theory" and "genre" point to different aspects of the written work. However interlaced their development may be, content and form can still be analyzed and discussed separately.[11]

11. An example may help to illustrate this. In his study of French higher education, Brockliss described how Cartesianism and Newtonianism successively replaced the traditional Aristotelian doctrines taught in university physics courses. Although each new doctrinal wave broke through against considerable resistance, what is conspicuous is how little the new

To illustrate how genres link ideas to their historical context, I shall examine the *Archiv für die Physiologie*, a journal founded in 1795 by Johann Christian Reil, a professor of medicine at the Prussian University of Halle. The *Archiv* appeared at a moment of turmoil in German medical and scientific circles, and its pages present a beautifully detailed miniature portrait of contemporary intellectual developments. Beyond merely recording what people thought, however, the *Archiv* itself changed in response to its environment. Not only did the ideas it published change over its lifetime, so too did the kinds of writings it contained. I propose to show here how those two developments were connected.

Physiology and Medical Writing in Late Eighteenth-Century Germany

In one sense, it is not surprising that Reil would choose to start a medical journal in 1795. The second half of the eighteenth century in German central Europe witnessed a remarkable thirst for reading on the part of the educated middle class, a phenomenon much commented on by contemporaries under such deprecations as *Lesesucht* and *Vielleserey*. While the production of books certainly soared during this period, even more prodigious was the growth in the number of new periodicals. In the 1780s alone, 1,225 periodicals were launched in German-speaking parts of Europe, roughly doubling the number of new titles produced during the previous decade. For the period 1765 to 1790 as a whole, a total of 2,191 new periodicals appeared, representing a threefold increase over the number of periodicals published in the previous quarter century.[12]

Along with other forms of literature, medical journals also enjoyed tremendous growth after midcentury. In the twenty-five years between 1766 and 1790 119 new titles appeared, compared with only 18 in the preceding quarter century. This pace accelerated in the 1790s, as 70 new periodicals

---

doctrines changed the curriculum. Once accepted, it turned out to be as easy to write a physics textbook, or a series of lectures, from a Cartesian or Newtonian standpoint as from an Aristotelian one. The three theories, at least as they were presented in the physics courses, can be seen as a group insofar as they shared the structural qualities of the genres in which they were presented. Nevertheless, they obviously said different things about the world. L.W.B. Brockliss, *French Higher Education in the Seventeenth and Eighteenth Centuries: A Cultural History* (Oxford: Clarendon Press, 1987), 350–370.

12. Joachim Kirchner, *Das Deutsche Zeitschriftenwesen: seine Geschichte und seine Probleme*, Teil I (Wiesbaden: Otto Harrassowitz, 1958), 72–73, 115.

were launched in that decade alone.[13] The overwhelming majority of these publications fell into two general categories. The first group consisted of journals of popular medical enlightenment, casting the warm glow of "rational" health care into the darkest corners of public ignorance. Explicitly didactic in tone, these journals aimed for a readership among the general public. The second category of periodicals was oriented more toward the community of health practitioners itself. Some of them contained nothing but reviews of new monographs, while others published digests and extracts from the foreign medical literature. Still others presented a hodgepodge of noteworthy case histories, new prescriptions, tips on keeping healthy, warnings against patent medicines and other forms of quackery, and essays covering a variety of subjects ranging from public health issues to the personal qualities most desirable in a doctor. Finally, a small but growing group of journals was published for practitioners in specialized areas such as public health, surgery, military surgery, obstetrics, and ophthalmology.

But if Reil's foray into medical journalism appears unexceptional from this perspective, his choice of subject matter was remarkable. The *Archiv für die Physiologie* was quite unlike any medical journal that had appeared to date in Germany, or indeed anywhere. Instead of offering material of interest to practitioners or their wealthier or more educated clients, Reil assigned to his journal the task of criticizing and advancing medical theory. Not only did this signal a major departure from the standard fare of medical periodicals, it also opened physiology—and medical theory in general—to a literary form in which it had never previously appeared. The placing of physiology in a *journal* opened the door to one of those new pairings between theory and genre of which I spoke above, and what made such a new pairing possible was the disruption of physiology's customary role in the constellation of eighteenth-century German science.

13. Ibid., 116. It should be noted that the large majority of these periodicals did not survive more than a few years, which makes the longevity of Reil's *Archiv* remarkable. Valuable publication information on all periodicals from the period is contained in Joachim Kirchner's *Bibliographie der Zeitschriften des deutschen Sprachgebietes,* 3 vols. (Stuttgart: Anton Hiersemann, 1969–1977). It is regrettable that little is known about the economics of publishing periodicals, a critical topic for the social history of ideas in eighteenth-century Germany, although it is clear from the huge number of periodicals begun during the period that launching a journal was not enormously risky for a publisher. For some discussion of the economics of publishing, see Hans-Martin Kirchner, "Wirtschaftliche Grundlagen des Zeitschriftenverlages im 19. Jahrhundert," in Joachim Kirchner, *Das Deutsche Zeitschriftenwesen* Teil II, 379–385; and Pamela Currie, "Moral Weeklies and the Reading Public in Germany, 1711–1750," *Oxford German Studies* 3 (1968): 74–75, 83.

In striking contrast to subjects such as chemistry, astronomy, and mechanics, physiology comprised one portion of a well-defined and highly structured program of professional instruction. The primary function of professional education in the German universities before the nineteenth century was the anointing of students with the marks of learnedness from which they drew their social status, rather than training them directly for the practice of their profession.[14] Guided by well-established tradition, each medical subject contributed its allotted portion to the universe of medical knowledge. Physiology played its role in two ways. First, it gave physicians a framework for interpreting the phenomena of life, often in concert with the leading philosophical currents of the day. Then, descending from the general to particulars, as was proper, physiology taught students the function of each organ in its turn. Hermann Boerhaave, the renowned medical teacher from Leyden, furnished the model for this kind of physiological writing in his *Institutiones medicae*, which became the basis for many subsequent eighteenth-century textbooks.[15] Because of physiology's specific function in medical education, a scholar wishing to make a name in this field did not have to enrich the store of knowledge by making empirical discoveries himself. It was just as reasonable to offer a new interpretation of the known processes of life, either by once again poring through standard authorities for new insights or by assimilating recent discoveries into the structure. The cardinal virtues of physiological writing—breadth of coverage, systematic ordering of materials, and force of argumentation—reflected its pedagogical setting.

As an academic subject, physiology had traditionally been presented in three major genres. Given its overwhelmingly pedagogical orientation, we should not be surprised that many professors published textbooks as accompaniments to their lectures on physiology. Scholars also produced comprehensive treatises dealing with particular topics or covering physiol-

14. R. Steven Turner, "The *Bildungsbürgertum* and the Learned Professions in Prussia: The Origins of a Class," *Social History/ Histoire Sociale* 13 (1980): 105–135; and Wilhelm Roeßler, *Die Entstehung des modernen Erziehungswesens in Deutschland* (Stuttgart: Kohlhammer, 1961), 95–125. One of the clearest expressions of this view of the medical profession was Friedrich Boerner's *Nachrichten von den vornehmsten Lebensumständen und Schriften jeztlebender berühmter Aerzte und Naturforscher in und um Teutschland*, 3 vols. (Wolfenbüttel: Johann Christoph Meißner, 1749–1756). Boerner, himself a physician, published detailed biographies of prominent physicians in German central Europe, along with lists of their scholarly and popular writings. In effect, Boerner wrote professional hagiographies of physicians emphasizing their educational accomplishments and their erudite credentials. The fact that these people were also healers rarely came up.

15. *Institutiones medicae, in usus annuae exercitationis domesticos digestae ab Hermanno Boerhaave.* Lugduni Batavorum, ex officina Boutesteniana, MDCCXXVII.

ogy as a whole. Although these two genres presented the highest profile in academic writing, a third and often neglected group of writings, Latin doctoral dissertations, also played an essential role in the creation of physiological theory. The publishing of dissertations allowed scholars to place limited portions of physiological theory under intense scrutiny and criticism without subjecting their ideas to the risks and rigors of a fully elaborated presentation. It is significant in this regard that many dissertations were not the work of the doctoral candidate himself. Instead, a dissertation was often written by a professor, then published at the student's expense with the student listed on the title page as the *Respondens* (that is, the person defending its theses in disputation). This arrangement brought appreciable benefits to both parties. It allowed professors to criticize existing doctrine, float new ideas, or report empirical observations, and it filled their pockets as well, since students paid a fee for the product. Meanwhile, the public defense of the theses during his graduation ceremony gave the student an opportunity to display his learnedness and rhetorical acuity.[16]

By the 1790s, however, the academic and professional environment for physiological teaching and writing had begun to change dramatically. New attitudes began to take shape concerning the relationship between the theoretical instruction offered by the universities in their faculties of law, theology, and medicine, and the practice of those professions in society. Critics assailed medical education for its lack of attention to preparing students for practice and for the mildewed pedantry of its teachers. In some quarters, proposals circulated for eliminating universities entirely, as the French had done, and replacing them with specialized academies that would offer a more suitable and practical type of instruction. The full measure of these reforms and their impact on German medical science must be taken elsewhere. For the present, I want to stress how greatly these criticisms, together with the curricular changes instituted in response to them, created uncertainty over the role of theoretical subjects such as physiology in the professional and educational structure of medicine.

16. Unfortunately, there has been little research on academic dissertations, and much of what there is is quite old. Ewald Horn, "Die Disputationen und Promotionen an den deutschen Universitäten vornehmlich seit dem 16. Jahrhundert," *Beihefte zum Centralblatt für Bibliothekswesen* 4 (1893–1894): 1–126. Also G. Kaufmann, "Zur Geschichte der academischen Grade und Disputationen," *Centralblatt für Bibliothekswesen* 11 (1894): 201–225. Fortunately, there has been a trace of a revival in recent years. Werner Kundert, *Katalog der Helmstedter juristischen Disputationen Programme und Reden 1574–1810*, Repertorien zur Erforschung der frühen Neuzeit, vol. 8 (Wiesbaden: Otto Harrassowitz, 1984), 53–75, gives an excellent description of both the forms and functions of academic dissertations and should be consulted as the latest—and certainly the best—word on the subject.

It was precisely to find a new mooring for physiology that Reil founded the *Archiv*. In the preface he wrote for its first issue, he complained about the backward state of physiology, placing the blame for this situation not on a lack of empirical knowledge about the body, but on methodological deficiencies in the way the subject had been approached in the past. "There lacks," he wrote, "an appropriate, prescribed plan, and also a knowledge of the rules by which we must investigate physiology."[17] The subject stood in great need of a thoroughgoing overhaul, and in the *Archiv* Reil took upon himself the task of organizing the project.

Reil's Crusade for a New Physiology, 1796–1804

The starting point for Reil's new theory—as for many German writers in the 1790s—was the epistemological theory of Kant's *Critique of Pure Reason*. Like Kant, Reil adopted a phenomenalistic approach to knowledge in "Von der Lebenskraft" (On the life force), the introductory essay in which he presented his theory. He began by arguing that the phenomena of sensation have their basis in matter, perceived as an object in space. The particular characteristics of these phenomena, he continued, arise from the properties of the matter at the origin of those phenomena, and especially from the form and mixture of that matter. If the phenomena change, so too must there have been a corresponding alteration in the form and mixture of the substance underlying them.[18] In attributing physiological processes to the "form and mixture" of matter, Reil expressed his conviction that the complex phenomena of life could be analyzed into simpler, more elementary chemical processes. Living matter, he conceded, may well have its own distinctive laws of chemical affinity, but he had not the slightest doubt that such laws, when established, would explain how life occurs.

For all Reil's ambitions in the realm of physiological theory, however, he never lost contact with the clinical and practice aspects of medicine. At a time when the relevance of medical theory to bedside practice was seriously being called into question, therefore, Reil attempted to place physiology at the base of a unified science of medicine. He applied his theory of life to the phenomena of pathology and therapeutics, arguing in "Von der Lebenskraft" and later essays that illness was simply the deviation of matter from

17. *Archiv für die Physiologie* 1 (1) (1795): 4.
18. Ibid., 14–15.

its proper form and mixture, and that therapy functioned to alleviate the problem by restoring the balance. Reil even went so far as to support the traditional and much-criticized division of his profession, between internal medicine and surgery, on the grounds that medicine cures by altering the mixture of matter with medicaments while surgery heals by restoring the body's healthy form.[19]

These were the main themes of the program that Reil articulated over the first several volumes of the *Archiv*. Although he presented it as a bold new departure, we should note that the *method* of inquiry—clearing away the clutter of hypotheses, establishing the proper principles, and building the system—treated physiology as it had long been treated in academic medicine. Reil envisioned physiology largely as a subject to be systematized in lectures and argued in disputation. Still, Reil's theory marked a significant shift away from the prevailing anatomically based theories of physiology, and perhaps its most novel feature was its clear articulation of a research program.

The question for us is, why did Reil choose a *journal* as the vehicle for his new physiology? Unfortunately, I have no privileged access to Reil's intentions in this matter, but he clearly wanted "Von der Lebenskraft" to establish a research program for physiology, and he wanted the journal to collect and record individual contributions to that program. By publishing a journal as opposed to a textbook, Reil emphasized the cooperative and programmatic nature of his new physiology. What is surprising—as well as characteristic of Reil's conception of physiology as a science—is the breadth of writings he counted as contributions to his program, a breadth that manifests a pronounced tension between Reil's theoretical aspirations, his choice of a journal as the vehicle for them, and the German audience for physiology at the end of the eighteenth century. Indeed, that breadth made the *Archiv* not so much a scholarly genre itself as a collection of genres drawn from other sources.

The first, and in many ways the most prominent, genre in the *Archiv*, the dissertation, was a direct carryover from earlier physiological writing. Because of their brevity and focus on one limited topic, dissertations were the

19. Reil alluded briefly to the nature of illness and therapy almost as an afterthought to "Von der Lebenskraft," 157–162. He returned to this question in two subsequent essays in the *Archiv*: "Ueber die nächste Ursache der Krankheiten," 2 (1797): 209–231; "Veränderte Mischung und Form der thierischen Materie, als Krankheit oder nächste Ursache Krankheitszufälle betrachtet," 3:424–461. For a more detailed description of Reil's theories, see Thomas H. Broman, "University Reform in Medical Thought at the End of the Eighteenth Century," *Osiris*, 2d series 5 (1989): 36–53.

one traditional scholarly genre suitable for a journal format. The disadvantage of dissertations had always been their exceedingly small press runs, usually below 200 copies, and their limited distribution. Intended originally for the audience attending the public defense, dissertations did not circulate through book dealers and book fairs as did monographic literature. Instead they were sent by post to correspondents, and they were commonly presented to princes and other patrons by those applying for jobs. Dissertations also became known through the various learned periodicals that sprouted everywhere during the eighteenth century. Universities used these publications, which were often edited by members of their faculties, to advertise themselves, and they invariably carried digests from local, recently defended dissertations.[20] Although these journals publicized certain dissertations, they did not translate and republish them. Thus the actual circulation of dissertations remained limited and largely informal, although prominent scholars occasionally published collections of them, as Albrecht von Haller did with an eight-volume series of anatomical dissertations in the 1740s.[21] The *Archiv* opened up more public and regular lines of distribution.

Beyond publishing German translations of his students' Latin dissertations, Reil adopted their rhetorical format to carry the burden of his theoretical argument. This was a task for which they were well suited, since from the time when they had been simple listings of theses to be handled in scholastic disputation, dissertations had gradually developed to a point where they displaced oral disputations from their central role in higher education. Of course, not every eighteenth-century dissertation was an exercise in scholarly argumentation. Some dissertations, especially those treating anatomical topics, reported the results of empirical investigations; others less memorable merely collected information on a topic from published sources. Many dissertations, however, continued to be written as a display of rhetorical skill. Even one of the eighteenth-century German names for it, *Streitschrift*, pointed to this argumentative quality.

These rhetorical devices were nicely displayed in "Ueber die Wirkungsart der Reize, und der thierischen Organe" (On the manner of action of stimuli and of animal organs), translated from the unfinished dissertation of

20. The most famous of these was the *Göttingische Gelehrte Anzeigen,* which was edited by Albrecht von Haller during his twenty-year tenure at Göttingen, and which continues to be published today. Other titles can be found in Kirchner, *Bibliographie der Zeitschriften* (see n. 13, above).

21. *Disputationum anatomicarum selectarum, collegit, edidit, prefatus est Albertus Haller.* Gottingae, A. Vandenhoeck, 1746–1752.

David von Madai, a medical student at Halle.[22] Madai's work dealt with the nature of organic stimuli, a subject much discussed by eighteenth-century physiologists following Haller's identification of sensibility and irritability as two fundamental properties of living matter. Madai presented his work as an investigation of how stimuli affect irritable parts of organisms, and how they produce reactions. As was typical, he began with a brief survey of what previous writers had said on the subject, a survey made largely without comment, suggesting that it served the ends of erudition rather than argumentation. Having disposed with this formality, Madai moved in the second section to the matter at hand, proclaiming it to be "the most important problem in all of medicine." After making a disclaimer about presenting only a "few ideas on this topic," Madai then set out his central claim: "Briefly, my opinion is the following. I believe that the actions of animated bodies and their individual organs are effects of change in mixture, which occurs at the same time as the actions."[23] This idea, which is obviously the same one Reil voiced in "Von der Lebenskraft," then became the foundation for Madai's subsequent explanation of how organs react to stimuli. Although Madai repeated much of Reil's system, he nonetheless developed it along lines that Reil himself had only touched on.

In his argument, Madai relied on making a rhetorical appeal to an audience holding common standards of reason and logic, amply demonstrated by his constant exhorting of his readers in the first person plural. He described how the world presents itself to "us," on what reasonable grounds "we" conclude this or that, and what "we" know or do not know. He offered no personal experience of the empirical world, and he described not a single experiment performed in pursuit of his questions. Instead, he applied the tools of logic and rhetoric to carry the reader along in a train of argument that seemed to move on a frictionless path toward its goal.

Madai made only two major departures from this style. In the third section, he recited a number of examples from inorganic nature to prove

22. *Arch. Physiol.* 1 (3) (1796): 68–148. Reil appended a note at the beginning of the article, explaining that Madai had died suddenly before defending the dissertation and that the current version had been prepared from the author's papers. In effect, Reil told the readers that he did not claim the work as his own, although he also did not disavow it, since he added that the dissertation merited publication for containing "so many ideas on one of the most important but also one of the most difficult physiological matters" (68).

23. Ibid., 82–83. In general, dissertations such as Madai's appeared to insert the author's voice into the discussion far more often than other types of writing, with the exception of experimental research reports. But the use of "I" in these two cases obviously is quite different. In the experimental report, "I" am telling the reader that things actually occurred, whereas in dissertations "I" am underscoring that what is being said is only "my" opinion, for which "I" take personal responsibility.

that changes in form and mixture of matter underlie the varying phenomena presented by nonliving objects. These examples were presented as straightforward, impersonal declarations of fact: "If a crystalized salt disintegrates into a powder, then it has previously been deprived of one of its components, namely its water of crystalization."[24] These facts then served to set up in the fourth section an analogy between the inorganic and organic worlds. "It is most probable," he wrote at the opening of that section, "that stimuli in the animate world work in a similar manner." In the very next paragraph, this presumption is slyly transformed by Madai into a fact. "There takes place in the stimulated organ a change in mixture."[25] The other departure occurred in the fifth section, where he paused to answer possible objections to his argument. Here too he abandoned authorial presence completely in order to allow the impersonal facts, or demonstrative logical arguments, to knock down the straw men he had set up.[26]

I have emphasized Madai's dissertation because this kind of argumentation, which seems to draw the reader into a circle of rational, like-minded individuals, figured centrally in the early volumes of the *Archiv*. Insofar as the journal's contents attempted to advance Reil's program, this rhetoric served Reil's purposes well. Reil's own essays, in which he set out the principles of his new physiology,[27] adopted virtually the same rhetoric, although he dispensed with the literature review that appeared to be mandatory in dissertations. Indeed, so tightly linked were Reil's essays with academic dissertations that in an introductory footnote to "Von der Lebenskraft," Reil declared the essay's main points to be elaborations upon material contained in three dissertations defended at Halle in 1793 and 1794.

Although Reil drew heavily upon dissertations as a model for his new venture, he tapped other genres of argumentation and criticism that existed in the periodical literature. One of these was the published letter. He sought to link medical theory to a broader intellectual movement that saw all scholarly disciplines, or *Wissenschaften*, as methodologically unified and capable of yielding fundamentally identical truths about the world. Consequently, Reil opened the journal's pages to philosophers to present their criticisms of existing medical theory and their suggestions concerning the requirements of a true *Wissenschaft* of medicine. Several of these appeared

---

24. Ibid., 88.
25. Ibid., 89–90.
26. Ibid., 93–95. Similar rhetorical tools are displayed in C[arl] A[rnold] Wilmans, "Ueber die Normalgesetze und ihren Nutzen in der Arzneykunde." *Arch. Physiol.* 5 (1802): 137–143. Wilmans article contained an excerpt from his dissertation, which was likewise defended at Halle.
27. See n. 19 above.

as relatively brief letters, praising Reil for how far he had advanced medical theory and containing "a few thoughts" on a particular matter.[28] One of the other philosophical contributions extended nearly to the length of a full treatise.[29]

There was still one further genre of criticism published in the *Archiv:* the book review. Like many of the most popular periodicals of the day, Reil's journal contained notices of recent publications that presented both summaries of their contents and criticism of their arguments.[30] These reviews gave Reil the opportunity to engage in a sort of dialogue with his contemporaries over questions of physiological theory, and he devoted considerable space to precisely those books that attempted to present comprehensive theories of life. The structure of these reviews was simple. Reil typically opened with fulsome words of praise for the inestimable contributions to science made by the writer, followed by a lengthy summary of the book's contents. When he wanted to insert brief objections to the author's ideas at a particular point—something he loved to do when there was an obvious disagreement with his own dogma—he adopted a technique widely practiced in other journals, that of placing his own remarks in parentheses, ending with "R." to designate the reviewer (*Rezensent*). These brief, critical insertions often took the form of questions, a clear, if also mild, way of expressing disagreement. For the more important reviews, Reil concluded with extensive remarks, explaining where he agreed with the author or, more typically, where the author had gone wrong. This explanation often amounted to a restatement of points from "Von der Lebenskraft" or one of his other major essays.[31]

It would do Reil a serious injustice to leave the impression that he was

28. See, for example, "Auszug eines Briefes des Herrn Professor C.C.E. Schmid zu Jena an den Professor Reil vom 9ten December 1797," *Arch. Physiol.* 3 (1798): 148.

29. Johann Köllner, "Prüfung der neuesten Bemühungen und Untersuchungen in der Bestimmung der organischen Kräfte, nach Grundsätzen der kritischen Philosophie," *Arch. Physiol.* 2 (1797): 240–396.

30. The peculiar affinity on the part of Germans for reviews and review journals in the late eighteenth century was discussed by David Kronick in his interesting study, *A History of Scientific and Technical Periodicals: The Origins and Development of the Scientific and Technical Press 1665–1790.* 2d ed. (Metuchen, N.J.: Scarecrow Press, 1976), 191. Kronick produced evidence showing that the Germans far outstripped all other nationalities in the production of review literature in the seventeenth and eighteenth centuries, and he joined previous writers in attributing this to Germany's rather meager role in the production of new scientific knowledge, which left German writers mostly to react to what others were doing. But I think this misses the point. It appears to me more likely that German scholars saw the type of criticism contained in reviews as setting the foundations for the systematic comprehension of knowledge they took as the highest standard of *Wissenschaft.*

31. See especially the reviews of J[oachim] D[ietrich] Brandis, *Versuch über die Lebenskraft,* 1(2) (1796): 178–192, and Christoph Heinrich Pfaff, *Ueber thierische Elektricität und Reizbarkeit,* 1 (3) (1796): 163–174.

some latter-day scholastic who believed the true physiology could be created solely from the reasonings of philosophers. He believed his reformed science would also require the discovery of new empirical knowledge, and for that reason he published one further genre in his journal: the empirical research reports characteristic of periodicals in the natural sciences. Reil had been deeply impressed by the advances made particularly in chemistry during the previous decade or so, and in "Von der Lebenskraft" he explicitly pointed to the analytical methods of the French chemists as holding the key to empirical studies of the form and mixture of matter. Reil's expectations for chemistry rapidly became apparent as he translated and reprinted in the early volumes a number of articles by the French chemists Fourcroy, Vauquelin, and Parmentier, containing results of their analyses of blood, tears, mucous, urine, liver, and brain. In a note attached to one of these articles, Reil praised the utility of chemical analyses and urged German chemists to take up these problems.[32] He also applied dissertations to this problem, publishing a number of those defended by his students at Halle which described the changes in form and mixture of various diseased or injured organs.[33]

Thus through its first five volumes the *Archiv* constituted a curious piece of work, mixing academic dissertations, empirical research reports, book reviews, and the essays of criticism and argumentation that were so characteristic of German periodical literature. However strange the blend, a large portion of it did seem to contribute, directly or indirectly, to Reil's plan for a unified medical science. Such programmatic unity in the early volumes is hardly surprising, since Reil obviously supplied a great many of these materials himself. The unity does suggest, however, that the goals described in "Von der Lebenskraft" have to be taken seriously as something more than prefatory bluster.

In one sense, Reil chose his tools well in attempting to make a new science of physiology. The various contributions to the *Archiv* both laid out a penetrating critique of existing physiology and indicated the methodological and empirical paths along which the subject might advance in the future. Both types of contributions in the *Archiv*, the critical and the empirical, were workable within the format of a periodical: critical contri-

32. Fourcroy and Vauquelin, "Zergliederung der Thränen und des Nasenschleims," *Arch. Physiol.* 1 (3) (1796): 52–53. This was but one of many chemical articles translated and published in the *Archiv*.

33. These dissertations were published in volume 4, pages 222–289, 365–387, and 387–412, and volume 5, pages 1–66. Unlike Madai's dissertation, they were not so much arguments for particular theses as they were repositories of pathological observations from various sources.

butions through the genres of the dissertation, philosophical letter, and book review, and empirical contributions through research reports.

Yet, precisely because it was in a journal, Reil could not maintain sole control over his program. He seems to have assumed his plans would enjoy the acquiescence and active participation of the *Archiv*'s readers and contributors. It soon became evident, however, that most other scholars had little interest in enlisting in Reil's program of a unified *Wissenschaft* of medicine. Guided by the interests and situation of this audience, the *Archiv* began to evolve in a direction quite distinct from the one envisioned by its founder. In the end, this transformation assured its survival, for if the *Archiv* had failed to match the needs and expectations of its audience, it could easily have shared the fate of scores of its contemporaries and vanished after a few issues. Instead, the *Archiv* did find its audience, but in doing so it ceased to be directed by Reil, developing instead its own voice and function in academic medicine. With a clearly defined function, the *Archiv* for the first time became a genre—a scientific journal—and not a collection of other genres. And at the same time, it began to establish a new pairing between theory and genre.

From a *Wissenschaft* of Medicine to Morphology: The *Archiv* After 1804

By 1805, when the *Archiv*'s sixth volume was published, the disintegration of Reil's program for a unified science of medicine appeared complete. Gone were translations of articles by the French chemists, the student dissertations recounting the form and mixture of diseased organs, and gone too were the book reviews and philosophical essays. In their place were published ever larger numbers of anatomical studies, some of which addressed the traditional functional concerns of physiology. The incompatability of Reil's private vision with the public and cooperative nature of a successful journal certainly played a major role in the program's demise, but such incompatability was not inevitable. Reil's program could well have corresponded to the vision of physiology held by his contemporaries. That it did not testifies perhaps in part to Reil's naivete, but also to the extraordinarily fluid situation that prevailed in German academic medicine at the opening of the nineteenth century. Academic physicians simply were unable to agree on what role physiology and physiological writing should play in medicine.

The instability of the academic environment for medicine in the years immediatley before 1800 is demonstrated by the rise of two movements, Brunonianism and *Naturphilosophie*, both of which posed direct challenges to Reil's program for a unified medicine. Brunonianism was a medical system that, like Reil's, proposed a link between the theoretical medical subjects and therapeutics. But unlike Reil's theory, which was put forth as a starting point for further criticism and empirical research, Brunonian physicians established a physiological model from which they claimed to draw clear consequences for the treatment of illness.[34] This challenged certain assumptions, long cherished by physicians, about the nature of bedside medical practice, and the ensuing uproar undoubtedly persuaded many writers to avoid the touchy issue of combining theory with practice.[35]

Naturphilosophie was quite another matter. It originated in the system of natural philosophy devised by Friedrich Schelling, in which Schelling used the properties of living organisms to elaborate a metaphysics radically divergent from the critical philosophy of Kant. Schelling's system attracted many enthusiastic admirers among younger physicians, who saw in *Naturphilosophie* the promise of unifying all human knowledge under the banner of philosophy and medicine, the academic home of the life sciences. This movement undercut Reil's program from the other direction, because most of its adherents disdained any interest in applying their theory to bedside practice. They saw themselves working toward something far grander.[36]

In quite different ways the two movements expressed deep dissatisfaction with the existing structure of medicine, both as it was practiced in society and as it was taught in universities. Of the two, however, it was *Naturphilosophie* that left its stamp on the theoretical medical subjects, especially physiology, more deeply. It offered scholars a new way of writing about the world, and it provided theorists with a discourse that allowed

34. Brunonian medicine has lately been the subject of a collection of essays assessing the movement in different parts of Europe, including Germany. W.F. Bynum and Roy Porter, eds., *Brunonianism in Britain and Europe, Medical History,* Supplement No. 8 (London: Wellcome Institute for the History of Medicine, 1988).

35. The notion of medical practice as an art based upon wisdom, judgment, and experience has appeared in diverse historical circumstances as an argument against attempts to make medical practice rigorously scientific. For example, see John Harley Warner, *The Therapeutic Perspective: Medical Practice, Knowledge, and Identity in America, 1820–1885* (Cambridge, Mass.: Harvard University Press, 1986), chaps. 2 and 3, and Christopher Lawrence, "Incommunicable Knowledge: Science, Technology and the Clinical Art in Britain, 1850–1914," *Journal of Contemporary History* 20 (1985): 503–520.

36. For a discussion of the doctrines of *Naturphilosophie* and their attraction for young physicians, see Thomas H. Broman, *The Transformation of Academic Medicine in Germany, 1780–1820* (Ph.D. diss., Princeton University, 1987) chap. 2.

them to assert their independence from the clinical side of medicine. It was not long before the *Archiv* began inclining in the direction of the new winds.

A two-year hiatus followed the *Archiv*'s sixth volume as French troops occupied Prussia and closed Halle, temporarily throwing Reil out of a job. When publication resumed in 1807, the *Archiv* had a new coeditor, Tübingen professor Johann Heinrich Autenrieth, and it had a new program: morphology, the study of animal form. Right from its first page, it was evident the *Archiv* had undergone some kind of transfiguration. "Under a theory of human anatomy," Autenrieth wrote to introduce an article on the anatomical forms of men and women, "I understand the doctrine of its laws of form [*Bildungsgesetze*]. . . . It is obvious," he continued shortly thereafter, "that here one cannot speak of the use of structures, but solely of their physical necessity."[37] Autenrieth then applied this principle in exhausting detail to argue that the differing bodily forms—ovoid for women and conical for men—derived from the opposing polar qualities of females and males. He pointed out that the smaller rib cage of women reduced the amount of air they could inhale, and this in turn gave women more negative, less oxidized qualities, such as greater sensitivity to external stimuli, and that men presented positive qualities, such as greater capacity for abstract thought. As generations of physiological writers had done before him, Autenrieth simply dressed up a collection of well-known "facts" in clothing provided by the reigning natural philosophy. But the clothing itself was new, as the emphasis on polar relationships indicates. *Naturphilosophie* had arrived in the *Archiv*.

Reil too was anxious to display the influence of the new ideas on his own thinking, and in the journal's next issue he used the principles of *Naturphilosophie* in an essay explaining the relationship between the sympathetic nervous system and the central nervous system.[38] The first third or so of the essay was occupied by a detailed anatomical description of the sympathetic nervous system, including its connections with the central nervous system. Reil used this description to emphasize how the structure of the sympathetic system was complete in itself and not dependent on the central nervous system, and he laid particular stress on the fact that each system

37. "Bemerkungen über die Verschiedenheit beyder Geschlechter und ihre Zeugungsorgane, als Beytrag zu einer Theorie der Anatomie," *Arch. Physiol.* 7 (1807): 1–139. The quoted material is from pages 1 and 2.

38. "Ueber die Eigenschaften des Ganglien-Systems und sein Verhältniß zum Cerebral-System," *Arch. Physiol.* 7 (1807): 189–254.

appeared to develop independently before birth. The reason for this description becomes clear when one reads Reil's explanation of the physiological roles of the two systems and their relationship to each other. The explanation depended on taking movement (*Bewegung*) and form (*Bildung*) as dialectically related manifestations of the same universal process. This process and its twin manifestations, Reil claimed, are represented at both levels of organic existence—the lower, or "vegetative," level and the higher, or "animal," level. He declared that at the lowest levels of life form dominates over movement, "and the unconscious Idea objectifies itself in structure." At the level of what Reil called "animality," however, movement emerges "ever more freely as visible and deliberate movement, and the Idea raises itself step-by-step toward consciousness."[39] Only in man is the final step of animality reached, with the result that vegetative and animal existence have emerged in their full dialectical glory. Thus humans have two parallel nervous systems, the sympathetic nervous system representing the vegetative, physical side of life, and the central nervous system manifesting the formal aspects of the animalistic, psychic side of life.[40]

In the limited space available here it is impossible to describe just how enormously different was the type of writing about nature offered in Reil's and Autenrieth's articles from that which the *Archiv* had contained previously. For the present, let me indicate three aspects of this difference. First, whereas previous scientific writing studied the properties of individual objects (corporeal or intellectual), *Naturphilosophie* proposed that the essential knowledge of the world lay not in individual objects at all, but rather in understanding the dynamic relationship between them. As we know, the relationships that most attracted attention were those between entities, such as the two poles of a magnet, which stood in opposition to each other and at the same time depended on each other for their existence. Second, what had previously been an *argument* about nature in terms of causes became instead in the discourse of *Naturphilosophie* an ordered *description*. Like the writings of his contemporaries, Reil's early writings aimed at finding the causes of phenomena (which were to be found in the form and mixture of matter), and whatever descriptions he gave were there to demonstrate the applicability of those causal explanations to the world. But in *Naturphilosophie* there was no search for causes, because all that needed to be known about the world was contained in the description of its

39. Ibid., 212.
40. Ibid., 213.

polarized or dialectical relationships. Finally, whereas physiological writing had previously aimed at elucidating the functions of organs, function and form in *Naturphilosophie* were elevated to the same "causal plane," with both representing the results of some more fundamental dialectical dynamism.

It may perhaps be surprising, but the essays by Reil and Autenrieth in volume 7 of the *Archiv* did not become the first in a long line of similar writings.[41] Rather, they served as theoretical prolegomena for a series of empirical investigations on anatomy. If the structures of the physical world truly were the external manifestation of noncorporeal processes, then anatomical study of the body could yield knowledge of the "physical necessity," as Autenrieth put it, by which the body is put together.[42] As Reil wrote at the beginning of an article on the anatomy of the cerebellum, "Reason in humans reflects itself in the organization of the nervous system, just as the Divine expresses itself in the corporeality of the entire cosmos."[43]

Thus the *Archiv* largely became a journal of anatomy and morphology, and what had been an abundant variety of articles shrank down to descriptions of anatomical structures and a scattering of theoretical essays and reports of chemical experiments on organic material. The *Archiv* still published translations of student dissertations, but even these looked substantially different from earlier ones. To begin with, each claimed to report the results of the student's own anatomical investigation. No longer did the *Archiv* publish dissertations that merely collected information from other sources. The topics of the dissertations were also new, covering subjects such as the spinal cord and spinal nerves.[44] And, of course, instead of framing an argument, these dissertations were structured around providing a description.

41. Reil did publish one more *naturphilosophisch* essay in the *Archiv* on the topic of pregnancy, "Ueber das polarische Auseinanderweichen der ursprünglichen Naturkräfte in der Gebärmutter zur Zeit der Schwangerschaft und deren Umtauschung zur Zeit der Geburt, als Beytrag zur Physiologie der Schwangerschaft," *Arch. Physiol.* 7 (1807): 402–501.

42. Timothy Lenoir, in *The Strategy of Life* (Dordrecht: D. Reidel, 1982), emphatically denied that *Naturphilosophie* provided the impetus for the early nineteenth-century German program of morphological research. Instead, Lenoir traced the foundation of this program to certain theories of Kant and the Göttingen professor Johann Friedrich Blumenbach concerning the teleological organization of life. While this may explain the origins of the program as a whole, certain aspects of it, such as Reil's and Autenrieth's adoption of morphology, can best be understood as deriving from the influence of *Naturphilosophie*.

43. Reil, "Fragmente über die Bildung des kleinen Gehirns in Menschen," *Arch. Physiol.* 8 (1807): 2.

44. G. G. Th. Keuffel, "Ueber das Rückenmark," *Arch. Physiol.* 10 (1811): 123–203, and Wilhelm Hermann Niemeyer, "Ueber den Ursprung des fünften Nervenpaares des Gehirns," *Arch. Physiol.* 11 (1812): 1–88.

**ity

A dissertation on the spinal cord written by a student named Keuffel illustrates the new orientation. After the still-mandatory review of previous writers on the topic, Keuffel organized his presentation of the spinal cord as one would see it sequentially during a dissection. First he described its sheaths or coverings, then the entire cord itself and its constituent substances (this being a morphological, not a chemical description), and then the internal structure of the cord.[45] Finally, Keuffel addressed himself to the origins of the spinal nerves. Although he followed the general order of dissection in his description of the spinal cord, Keuffel did not present his results as a *narrative* of the dissection.[46] Rather, anatomical structures were treated in terms of their position and appearance. Because Keuffel used his dissertation to present a series of facts rather than to construct an argument, the authorial voice only occasionally made its presence felt.

The author's voice did intrude, however, at two significant points. The first of these was where Keuffel wanted to make clear just which things *he* had discovered. To this end he brought earlier writers into the discussion to assess what they had said about certain structures, and then compared his findings against their claims. "Concerning the arachnoidea [a membrane that surrounds nervous tissue]," he wrote in one place, "I have nothing to add to what is known."[47] Similarly, he reported that "up till now" anatomists had believed the spinal cord always ended in the same location on the spinal column; however, he said, "According to my observations it is not always the same."[48] The second point at which Keuffel brought himself into the presentation was where he described a new technique he had developed to prepare the spinal cord for study. Once again, in part, this was to ensure that he received credit for his contributions. But Keuffel also used the first person to construct an account of the technique similar to the narrative of method contained in reports of experiments.

> I took a piece of spinal cord several inches long from a horse or ox in the region of the upper spinal vertebrae, where the smallest and fewest nerves emerge from the cord, and [I] laid it in a solution of caustic alkali in distilled water, of which one ounce contained one-half to one dram of alkali.[49]

With the exception of its historical survey, Keuffel's article was typical of

45. G. G. Th. Keuffel, "Ueber das Rückenmark," 136–179.

46. Indeed, as Reil pointed out elsewhere, such a narrative could not be given, since any single dissection necessarily destroys certain parts in order to be able to study the structure of others. Consequently, the description presented at the end is a composite of several individual dissections. "Fragmente über die Bildung des kleinen Gehirns in Menschen," 17.

47. Keuffel, "Ueber das Rückenmark," 137.

48. Ibid., 146.

49. Ibid., 163.

the *Archiv*'s contributions after volume 7. Physiology had indeed become a new science, though clearly not the one Reil had originally intended. The change in theoretical content is evident, from the chemical theory described by Reil in "Von der Lebenskraft" to the anatomical descriptions of morphology. This is not to claim that the link between anatomy and physiology was novel in this period; quite to the contrary, their association is an ancient one. But their relationship was new nonetheless. Previously, physiological function had provided the final cause of anatomical form, or form had furnished the formal or efficient causes of function, depending on the interests of individual writers. After 1800 *Naturphilosophie* provided the theoretical framework for examination of organic form for its own sake, as the external manifestation of physiological process. From a science of argumentation about causes, physiology in the thrall of *Naturphilosophie* became a science of description.

Alongside this theoretical transformation, and indeed inseparable from it, there developed a new way of writing about physiology. To put it most simply, physiological writing in German Europe began to lose its intimate connection with medical pedagogy. The rhetorical tools for doing physiology ceased to be those of the disputation and the lecture, a change readily seen by comparing the dissertations of Madai and Keuffel. Although the same name pertains to both, and at one level the writing of a dissertation continued to certify professional competence, the nature of that competence changed dramatically. Madai's dissertation retained a close link with established educational practices; we can easily imagine him defending its theses in his public disputation, had he lived to do so. We can also imagine him applying those same skills to the classroom himself someday, a task for which he would have been well prepared by his education. For Madai, being a doctor of medicine still meant, at least in part, that he was a teacher, a meaning preserved from the original Latin meaning of "doctor."

Keuffel's dissertation, by contrast, has little of the classroom about it. Although he too endured the formality of an oral disputation, his dissertation was not composed specifically with that in mind, and we can scarcely imagine what kind of argumentative give-and-take it could have engendered. Instead of certifying his ability to teach, Keuffel's dissertation certified his ability to discover things about nature. He wrote in an era when teaching and research had begun to emerge as distinct scholarly functions.[50]

50. There *was* one genre of classroom writing to which Keuffel's demonstration bore a close resemblance: the anatomical demonstration. Yet precisely because they were *demonstrations* and not *arguments,* such writings typically employed different rhetorical tools than those

Just as it influenced the kinds of medical dissertations being written, the separation of teaching and research created a specific niche for the *Archiv*. At a time when physiology had been dominated by genres of writing linked to pedagogy, the *Archiv* as a periodical lacked a clear function. With the recasting of physiology under the research program of morphology, however, the journal acquired a clear function in pursuit of that science, becoming the repository for individual contributions to morphological knowledge. No longer a motley collection of other genres, the *Archiv* was transformed into an empirical research journal.

The *Deutsches Archiv für die Physiologie* and the Community of German Morphologists

Reil's death in 1813 threw the future of the *Archiv* into doubt. In 1815, Autenrieth brought out one more volume, consisting largely of translations of foreign articles, but he appeared to be unwilling to assume responsibility for its continuation. Fortunately, by the time the final volume of Reil's *Archiv* was published, a new journal was already in the works: the *Deutsches Archiv für die Physiologie*, to be edited by Johann Friedrich Meckel, a former student of Reil's and, like his mentor, a professor at Halle. Meckel's journal was no mere continuation of Reil's publication, and several features marked it as a departure from the original. It came in a new format of four issues per year, compared with the three issues of its predecessor. The *Deutsches Archiv* was also larger and more richly produced than Reil's, each issue containing several engraved plates, some even hand colored, in contrast to the sparser illustration that had characterized the earlier *Archiv*.[51] The contents differed as well; each issue consisted of two sections, one for the publication of original research reports and the other for short digests of important physiological work published elsewhere. The *Deutsches Archiv* contained no book reviews and only a handful of general theoretical essays, conforming to its editor's stated goal of publishing a journal of empirical research.

---

dealing with physiology. Andrew Wear's study of Harvey's *De Motu Cordis* (see n. 2), in fact, makes a strong argument that Harvey treated the circulation of the blood in a manner like any other anatomical demonstration, and not as a physiological argument. Whatever their use in teaching anatomy, such demonstrative presentations had not been applied in lecturing on the theoretical medical subjects.

51. Not surprisingly, the number of engravings in Reil's *Archiv* had increased noticeably when it began publishing large numbers of anatomical contributions. Nevertheless, the *Deutsches Archiv* contained even more.

Of all these changes, however, probably most significant was the claim made on its title page that Meckel was editing the journal in conjunction with a group of other scholars, who were named on the same page. Whereas Reil, and later Reil and Autenrieth, had made no attempt to present their journal as a cooperative enterprise, Meckel openly asserted the existence of a community of physiologists who shared responsibility for the *Deutsches Archiv*. Nor was it any ordinary crew whom Meckel designated as his associates. They included such luminaries as Johann Friedrich Blumenbach, Ignaz Döllinger, Friedrich Tiedemann, Karl Friedrich Burdach, Carl Gustav Carus, and Karl Ernst von Baer—in short, nearly everyone who counted in German physiology during the period. Of course, we might doubt whether they actually ever performed any editorial services that would warrant their inclusion in that elite group. Be that as it may, Meckel's *Deutsches Archiv* presented itself as the organ of a self-conscious scholarly community in a way that Reil had never conceived. Far from declaring an independent science of physiology, Reil, as we have seen, had originally intended exactly the opposite. The story of Meckel and his journal, however, resembles the one told by Karl Hufbauer about the German chemical community in the eighteenth century. Hufbauer argued that a self-conscious community of chemists coalesced around the various journalistic enterprises of the indefatigable Lorenz Crell. Crell's chemical journals, according to Hufbauer, gave German chemists a forum for dialogue among themselves, instead of speaking primarily to foreign chemists or other local intellectuals. Clearly, the *Deutsches Archiv* wanted to present itself as doing the same thing for German physiologists.[52]

The appearance of this community can be understood against the background of events I have already described. In the first place, the separation of teaching and research functions occurring in the years after 1800 allowed writers on theoretical subjects such as physiology to define a set of specialized problems to be investigated by recognized experts. In addition, Reil's *Archiv* had already paved the way for specialists in physiology. The journal's appearance created a space in the existing forms of medical writing that could be exploited by morphologists. Then too, the doctrines of Schelling's *Naturphilosophie* suggested a set of research problems that might profitably be exploited. To these factors must finally be added one new development. After 1800, and especially after 1815, German medical faculties

52. Karl Hufbauer, *The Formation of the German Chemical Community* (Berkeley: University of California Press, 1982).

expanded in size and grew more specialized in their assignment of professorial responsibilities. Meckel's own case is illustrative in this respect. When he assumed his father's former chair at Halle in 1808, consisting of the subjects of anatomy, physiological anatomy, obstetrics, and surgery, Meckel elected to give up responsibility for the latter two clinical subjects in order to concentrate on anatomical teaching and physiological research. This example was but one case of a trend that saw anatomy pass from its eighteenth-century alliance with surgery to a new connection with physiology in the early nineteenth century. At many schools the new positions furthered the separation of clinical from theoretical subjects, a development that was also advanced by the acrimony surrounding Brunonianism from 1795 to 1805.[53] In its aftermath, clinicians wanted nothing more than that theoreticians stay away from the problems of medical practice. Holders of chairs in anatomy, who now customarily taught physiology as well, thus were free to pursue whatever research subjects they wished, as long as they continued to offer lectures on human anatomy to medical students.

Under these conditions, a community of researchers formed, and it constituted the outstanding characteristic by which the *Deutsches Archiv* was distinguished from its predecessor. The signs of the community's existence are to be found everywhere in the journal. Take for example the issue of priority. Priority had not been a major concern in the early *Archiv*, with its book reviews and philosophical essays. But once the dominant genre had become one of empirical research, the question of priority became a potentially contentious one. We have already seen hints of that concern in Keuffel's dissertation on the spinal cord. If he was not going to distinguish himself by constructing an elegant argument on a certain thesis, Keuffel spared no pains in ensuring that he received proper credit for his original contributions.

Once a community forms, the desire for recognition of priority in producing the first description of a certain structure becomes that much more pronounced, and in the *Deutsches Archiv* the concern for claiming priority for oneself and awarding it to others is incessant. Although we do not find any disputes over priority, we see concern for it manifested in authors' innumerable references to the time at which they first started doing research on a topic, which was always before somebody else published something on it. Thus one writer talked of reading about a certain observation in a monograph on the ligaments of the uterus, which reminded him

53. On Brunonianism, see above, p. 30.

that he had noticed the same thing while doing some dissections "several years ago."[54] Meckel in particular was a fiend about this. In the very first article published in his new journal, he noted that a number of monographs had recently appeared on his topic, the embryological development of the central nervous system. But, Meckel stressed, his own work had been undertaken independently of other research, and he supplied a date to lend weight to the claim.[55] In another instance Meckel attached an introductory footnote to an article by another prominent scholar, Friedrich Tiedemann, on the appendix in amphibians, saying that Tiedemann's article was received by him after his own article on the gut in amphibians (which appeared in a previous issue) "had been *printed,* but the issue *had not yet been distributed.*"[56] Meckel added that he immediately sent printer's copies of the pages to Tiedemann to prove that he had not stolen his ideas.

Besides priority, the impact of the community on the *Deutsches Archiv* is also evident in the structure of the conversation that ensued in it. In its early years Reil's *Archiv* had contained a variety of conversations. A few were running dialogues between two contributors on a particular subject; significantly, these conversations had nothing to do with Reil's program. Such conversations that did pertain to the program tended, not surprisingly, to be radial in structure. Contributors talked to Reil in letters and in certain essays, and Reil talked to others in book reviews. When we open the *Deutsches Archiv,* however, we find an active multiparty conversation. The contributors refer to each other's work constantly—acknowledging and rejecting discoveries claimed by others, accepting and rejecting their interpretations of the meaning of structures.

In two senses, in fact, the conversations between scholars published in the *Deutsches Archiv* quite literally *were* the community. First, in a nation as large and politically divided as Germany, lacking as it did a national capital, the journal provided the only suitable place for widely scattered scholars to meet with each other. Second, the individual claims about discoveries and judgments made about them constituted the core of the program followed by many of the journal's contributors. Morphology was a science that sought to elucidate general laws of form of living organisms. Although it did this under the guidance of certain theoretical principles, most mor-

---

54. Christian Ludwig Nitzsch, "Ueber die vordern runden Mutterbänder in Säugthieren," *Deutsches Archiv für die Physiologie* 2 (1816): 592.

55. "Versuch einer Entwicklungsgeschichte der Centraltheile des Nervensystems in den Säugthieren," *Deutsches Arch.* 1 (1815): 1–2.

56. *Deutsches Arch.* 3 (1817): 368. The emphasis was Meckel's.

phologists believed their ends could best be achieved through accumula-tion of empirical studies. And the major center for collecting those studies was *Deutsches Archiv*.

Over the twenty-five-year span between 1795 and 1820, the *Archiv für die Physiologie* and the *Deutsches Archiv* represented the evolution of German physiology at two levels. At one level it is evident that the role of physiology and physiological theory within medicine underwent a major transforma-tion. Reil's intention at the outset had been to establish firm epistemologi-cal foundations for a physiological theory that would link the diverse subjects of medicine, both clinical and theoretical. During the next decade the links between physiology and other areas of medicine, and to the larger university environment, withered as German physiologists defined a set of problems internal to their own subject and asserted their right to do research without paying heed to its potential clinical ramifications. Keuffel, in the introduction to his dissertation, expressed this forthrightly. He argued that anatomists had long avoided studying the spinal cord because they did not see how it could contribute to medical therapy. "But toward the end of the century just passed," he went on, "the correct ideas concern-ing life in general and its universal existence . . . came into more general circulation," and this provided the impetus for a new kind of anatomical study.[57] Several years later, in 1817, Meckel made a similar point. "The time will soon be gone," he wrote, "when one sees in [anatomy] only a mindless topography, and when one uses every attempt to reach Generality from the Particular only as the occasion for petty derision."[58]

A parallel change is seen in the evolution of the articles published in the journal. When Reil began editing the *Archiv*, he published in it a diverse range of material, much of which was relevant to his goals, but which also left the *Archiv* a patchwork of other scholarly genres rather than any one distinctly "periodical" genre. By the time Meckel assumed editorship, how-ever, the diversity of the *Archiv*'s offerings had diminished markedly, and its role in the construction of physiological knowledge had become more defined. As the medium of communication for a recognizable community of scholars, the *Deutsches Archiv* acquired a clearly defined function: to provide empirical building blocks of theoretical structures that would ap-pear in longer treatises.

Most important, the evolution of the *Archiv* into the *Deutsches Archiv* represents the establishment of a new pairing between theory and genre in

57. Keuffel, "Ueber das Rückenmark," 124. The "correct ideas" Keuffel alluded to are those of *Naturphilosophie*.
58. *Deutsches Arch.* 3 (1817): 396.

physiology. We have seen how the *Archiv* first appeared at a time when the content and function of medical theory had come under serious criticism, causing previously existing pairings between theory and genre to become loosened. Reil originally had attempted more to adapt his journal to the traditional pedagogical goals of academic writing than to break away from them. Yet despite his intentions, his journal soon began to change into something quite different in response to a new environment that existed for the theoretical medical subjects in Germany. The change that began in Reil's *Archiv* culminated in the *Deutsches Archiv*, which embodied the science of morphology and closed a new linkage between theory and genre.

This essay has argued that the appearance of certain new theories, such as morphology, can be explained as responses to the changing environments in which they are articulated, and that the social functions of literary genres mediates the relationship between theory and environment. As a conclusion, I would like to address two problems raised by this interpretation, the first of which concerns the generality of the claims. By no means do I want to say that the appearance of new scientific theories must always follow upon the kind of loosening between theory/genre pairings described here. Nor must they always be accompanied by the development of new genres of scholarly writing. Other mechanisms could well operate, although whatever their source new theories as *writings* would still have to serve the functions of the genres through which they are presented. At the same time, I do not believe the *Archiv für die Physiologie* was a unique case. There is good reason to believe a similar treatment could be applied elsewhere in the history of scientific ideas.[59]

Yet, in restricting the generality of this interpretation, a serious difficulty arises. Assuming that new theories can be generated from other sources and can be written about in existing genres, what becomes of the "pairing" of which so much as been made? This question recalls a point raised earlier in the essay:[60] if the structuring of contents (theories) by genres is not reasonably restrictive, then the utility of genres as mediators between theories and historical environment disappears. Genres would turn out to be merely convenient vehicles for presentation of ideas, retaining their social functions but having only a circumstantial relationship to the information they present.

Earlier I suggested that scholarly genres develop not in tandem with

59. Indeed, Hannaway's *The Chemists and the Word* (see n. 2) could be read as making just this kind of argument.

60. See p. 17.

individual theories, but with groups of theories. Some new theories, then, might be generated without disrupting existing pairings, because they are sufficiently similar to previous theories to be written about in the same genres. But what defines "sufficiently similar," and what criteria can be used to identify such groups of theories? In this essay I have tried to suggest that the purposes and content of morphology were so divergent from the type of physiology described by Reil in "Von der Lebenskraft" that it could have been articulated *only* through other genres. This suggestion should now be further developed through close reading to determine how the writings are organized and argued. Only when these structural elements are described will we be in a position to assess their relevance to the theories presented in them and to engage the questions raised here about the relationship between theory and genre.

Research for this article was supported by a postdoctoral fellowship in History of Science from the National Science Foundation. The author thanks the National Science Foundation and the Section of the History of Medicine at Yale University.

Lynn K. Nyhart

2. Writing Zoologically: The *Zeitschrift für wissenschaftliche Zoologie* and the Zoological Community in Late Nineteenth-Century Germany

ALTHOUGH THEY NOW CONSTITUTE the most important mode of published scientific communication, we know comparatively little about the history of scientific journals. In particular, the ways in which journals embody the social dynamics of the particular communities they serve remain insufficiently analyzed.

This lack may come as a surprise to readers familiar with traditional sociology of science. Certainly Merton, Zuckerman, Crane, and others have shown us that journals serve such functions as rapid communication, the establishment of priority claims, and "gatekeeping" (control of access to the opportunity to publish, and hence to recognition and reward).[1] Yet, without further investigation, there is little reason to believe that the editorial practices through which these functions are expressed in twentieth-century scientific journals are the same as those of earlier journals, or even that the functions themselves have carried the same weight in all scientific communities. Despite modern commentators' repeated references to the establishment of a rudimentary "editorial peer review" system in the early

1. Robert K. Merton, *The Scoiology of Science: Theoretical and Empirical Investigations,* ed. Norman W. Storer (Chicago: University of Chicago Press, 1973) reproduces many of Merton's most important papers, including Harriet Zuckerman and Merton, "Institutionalized Patterns of Evaluation in Science," 460–496, originally published as "Patterns of Evaluation in Science: Institutionalisation, Structure and Functions of the Referee System," *Minerva* 9 (1971): 66–100. See also Diana Crane, "The Gatekeepers of Science: Some Factors Affecting the Selection of Articles for Scientific Journals," *American Sociologist* 2 (1967): 195–201. The more recent social analysts of scientific knowledge, concerned mainly to show that scientific ideas and not just behaviors are constructed through social interactions, have done little to modify these basic precepts about behavior. An important exception is David Hull, who has developed an evolutionary model that accounts for behavioral patterns among scientists. Hull, *Science as a Process: An Evolutionary Account of the Social and Conceptual Development of Science* (Chicago: University of Chicago Press, 1988).

Philosophical Transactions of the Royal Society, for example,[2] we cannot assume that this has been the "normal" system of assuring the quality of scientific articles over the last three hundred years. Similarly, although the *Philosophical Transactions* set a precedent in using rapid publication to help assure credit for priority claims, we should not expect that the urgency of establishing priority or the ways in which priority disputes are dealt with have been the same in all scientific communities.

Historical discussions suggest that in fact journals have adopted quite a variety of practices expressing different relationships between editors, contributors, and readers, and that journals can take on a range of roles for their communities. The process of peer review for the *Physical Review* in the well-established, highly competitive international community of twentieth-century physicists is a complex affair reflecting, among other things, the size and stratification of the community and the journal's own standing as the most prestigious among many outlets for publication.[3] This hardly involves the same dynamics as those put into play by the leader of a nineteenth-century German research school such as Justus Liebig, who used his editorship of the *Annalen der Chemie* to promote aggressively a particular scientific program and the careers of those who supported it.[4] Both of these situations differ from that of the eighteenth-century editor Lorenz Crell, around whose journal *Chemische Annalen* a newly forming chemical community coalesced. Not only was the sense of community new and fragile,

2. Merton and Zuckerman, "Patterns of Evaluation," 462–470; Frank Manheim, "The Scientific Referee," *IEEE Transactions on Professional Communication* PC-18 (3) (Sept. 1975): 190–195, reprinted in *The Scientific Journal,* ed. A. J. Meadows (London: Aslib, 1979), 99–103; David Hull, *Science as a Process,* 324. Charles Bazerman points to a related "modern" device for quality control established by the *Philosophical Transactions* in the 1750s: the editorial board. Bazerman, *Shaping Written Knowledge: The Genre and Activity of the Experimental Article in Science* (Madison: University of Wisconsin Press, 1988), 137. Merton and Zuckerman, 469, show that the French *Journal des Savans* had already established a similar practice by 1702.

3. On refereeing in the *Physical Review,* see Merton and Zuckerman. Other studies of the referee system in particular twentieth-century journals include David Hull, *Science as a Process,* chap. 9 (on the journal *Systematic Taxonomy*), and Richard D. Whitley, "The Operation of Science Journals: Two Case Studies in British Social Science," *Sociological Review* 18 n.s. (1970) 241–258. Diana Crane, "The Gatekeepers of Science," cites earlier literature on refereeing.

4. J. P. Phillips, "Liebig and Kolbe, Critical Editors," *Chymia* 11 (1966): 89–98, gives a good sense of how Liebig used his editorial position to attack opponents of his program. The importance to a research school of having a ready publication outlet has been emphasized by J. B. Morrell in "The Chemist Breeders: The Research Schools of Justus Liebig and Thomas Thomson," *Ambix* 19 (1972): 1–46, and has since been taken as one of its common features. See Gerald L. Geison, "Scientific Change, Emerging Specialties, and Research Schools," *History of Science* 19 (1981): 20–40, especially Chart II. Using a somewhat different analytical model, Terry N. Clark has suggested that a journal may even function as a "research institute" when another institutional form is not available: "The Structure and Functions of a Research Institute: the *Année Sociologique,*" *Archives Européennes de Sociologie* 9 (1968): 72–91.

but even the concept of a disciplinary research journal was still novel. These circumstances imposed a set of demands on Crell different from those on editors who succeeded him.[5] Furthermore, as Thomas Broman's article in this collection shows, even a single journal can go through enormous transformations as the character of the community it serves changes.[6]

As these examples and their comparison suggest, historians of science still stand to gain by examining carefully scientific journals, not just for the contents of their articles, but as vehicles through which the social dynamics of a given scientific community are expressed. In addition, because contemporaries could not help but recognize that journals were communal enterprises, investigation of their journals allows us insight into the very concept of "scientific community" held by the past scientists themselves. This ought to be a crucial element in constructing a social history of science, and journals offer one way of getting at it.

This paper examines the social role of journals and the writings in them for the community of university zoologists in late nineteenth-century Germany. I begin by looking at the *Zeitschrift für wissenschaftliche Zoologie*, the leading zoological journal through the 1860s and 1870s. To its chief editor, Carl Theodor Ernst von Siebold (1804–1885), the *Zeitschrift* was to be a public repository of scientific research—facts and theories, but especially facts. At the same time, however, he also saw it as the embodiment of a community of scientists engaged in a joint enterprise. This aspect of the journal is invisible in its printed pages, but clearly evident in Siebold's

5. The authoritative source on Crell and the chemical community is Karl Hufbauer, *The Formation of the Chemical Community (1720–1795)* (Berkeley: University of California Press, 1982). Although the dynamics of a research journal for an entire discipline are not exactly the same as those of a particular research school's journal, in both cases the founders of such journals usually start them to promote a particular vision of their science. Compare the literature on research schools in fn 4 above with, for example, John Servos, "A Disciplinary Program That Failed: Wilder D. Bancroft and the *Journal of Physical Chemistry*, 1896–1933," *Isis* 73 (1982): 207–232.

These examples, of course, do not begin to cover the range of historical analyses of scientific periodicals. These run from the straightforward analysis of contributors, contents, and intellectual trends, as in Christa Jungnickel and Russell McCormmach's treatment of Poggendorff's *Annalen* in their *Intellectual Mastery of Nature: Theoretical Physics from Ohm to Einstein* (Chicago: University of Chicago Press, 1986), to a much greater focus on the dynamics of publishing itself, as in Susan Sheets-Pyenson's "From the North to Red Lion Court: The Creation and Early Years of the *Annals of Natural History*," *Archives of Natural History* 10 (1981): 221–249.

6. Other analyses of individual journals also indicate the extent to which change in a community is reflected in its leading journal: see, for example, David Hull's analysis of *Systematic Zoology* in his *Science as a Process*, or, for a very different kind of journal and community, Roy MacLeod and Paul Gary Werskey on the history of *Nature*: "Is It Safe to Look Back?" *Nature* 224 (1969): 417–476.

editorial correspondence. Not only was this sense of community central to his conception of the journal, but the very *invisibility* of its management, the firm boundary between the social activity of dealing with the contributors and the scientific matter actually printed in the journal, was intrinsic to his view of community and of science itself.

As I go on to show, in the 1870s and 1880s Siebold's conception of community, with its clear demarcation between public and private—between the objective results of science and the often messy process of achieving those results—lost favor as the zoological community itself outgrew a single journal and fissured into numerous competing and often mutually hostile camps. In the new atmosphere the sorts of conflicts Siebold sought to deal with behind the scenes found their way into print. The appearance of new journals that aired these disputes signaled more than just a period of growth and diversification of research programs (though this was certainly the case): by bringing these disputes, most of them over priority, into the printed pages of journals, it sanctioned a new code of scientific behavior. Just as the invisibility of social negotiations was central to Siebold's concept of the community, the public nature of the disputes in the late 1870s and 1880s betokened a very different sort of community, one in which competition and suspicion rather than cooperation and trust became the accepted norm. Yet as the final section suggests, the priority disputes themselves may be understood as more than simply the consequences of heightened competitiveness. Ironically, they provided a means for scientists in this new atmosphere of public display to declare their allegiance to the traditional norms of community behavior.

The *Zeitschrift für wissenschaftliche Zoologie*

The *Zeitschrift für wissenschaftliche Zoologie* was founded in 1848 by Albert von Kölliker (1817–1905) and Carl Theodor Ernst von Siebold. As initially planned in 1847, the *Zeitschrift* was to encompass botany as well as zoology, with Alexander Braun and Carl Naegeli editing the botanical section.[7] (Siebold and Braun were colleagues at the University of Freiburg at the time; Kölliker and Naegeli were at Zürich.) The title they chose for their

---

7. The new journal, originally intended to be titled the *Zeitschrift für wissenschaftliche Zoologie und Botanik,* was modeled on the short-lived *Zeitschrift für wissenschaftliche Botanik* edited by Naegeli and Jakob Schleiden, which went through only one volume (four issues) between 1844 and 1846.

journal was intended to express their program and convey an implicit critique of the dominant trend in natural historical work, which was systematics. Sidestepping the endless disputes over what arbitrary characters should be used to assign organisms to which taxonomic categories, their journal was designed to advance a truly *scientific* zoology and botany. The prospectus for the new journal drafted by Siebold and Braun and printed in early 1848 clearly states their intentions.

> We desire to give our journal the most scientific character possible. . . . in the objective sense of real scientific research, through the most comprehensive and pure [*geläuterte*] presentation of the facts, of their lawful determination and of their causal connection. To this purpose we exclude all announcements of new genera and species that do not relate to this task, unless these offer us a more thorough-going insight into plant and animal construction [*Bau*], into the life-history of animals and plants, or in the lawful organization of the organic realms. For the same reason we will exclude any kind of simple notes and natural history news. . . . On the other hand, from the truly scientific side of botany and zoology nothing will be excluded.[8]

"Scientific zoology," to the editors, meant several things. It meant treating a certain set of problems, namely those revolving around comparative anatomy, embryology, and physiology, rather than systematics; it also meant pursuing a certain set of aims, namely discovering the laws of living nature and their causes, by means of the general method of uncovering numerous facts and discovering the patterns that connected them. As we will see, this emphasis on empirical research was accompanied by a set of expectations about the scientific behavior of contributors to the journal: not only would the successful manuscript submission follow this approach to knowledge, but in doing so its author would avoid the pitfalls of placing personal vanity or opinions ahead of the facts.

Although the botanical half of the journal never appeared, the zoological half managed to survive the vicissitudes of 1848, and its editors went on to establish the *Zeitschrift für wissenschaftlich Zoologie* as a central organ for communicating research in the German zoological community. In fact, with two exceptions, all the zoologists who held university professorships in Germany in 1880 had contributed to it at some time in their careers.[9] As

8. Quoted in E. Ehlers, "Carl Theodor Ernst von Siebold. Eine biographische Skizze," in *Zeitschrift für wissenschaftliche Zoologie* (hereafter ZwZ) 42 (1885): i–xxiii, on page xiii.
9. The two exceptions, Alexander Goette of Strasburg and Theodor Eimer of Tübingen, regularly published in the *Archiv für mikroskopische Anatomie*.

Germany became the center of scientific life in the latter part of the century, the *Zeitschrift* came to international prominence, publishing works (in German) by authors from throughout continental Europe, Russia, England, and even the United States. (French authors are notably, and predictably, absent.)

The journal's rise to prominence was made easier by its finding an open niche in the publishing market—until the mid-1870s someone wanting to publish a zoological work had few journals to choose from. One likely candidate, especially for a person working on invertebrates, was the *Archiv für Naturgeschichte*, the only zoology journal older than the *Zeitschrift für wissenschaftliche Zoologie* that continued through the period under discussion. The *Archiv* was mainly devoted to systematics, however—exactly the area that the *Zeitschrift* disdained.[10] By the 1870s the *Archiv* may have had a reputation for being somewhat old-fashioned; it certainly was not publishing articles by many well-known people. Siebold once advised his managing editor that the best way to soothe an author whose manuscript he did not want was to tell him that the *Zeitschrift* already had too many manuscripts and to suggest that the author send his piece to the *Archiv*, "which is always short of manuscripts."[11]

A number of other journals representing various research programs bordering on zoology were also available to authors before 1870. One might try the *Archiv für Anatomie, Physiologie und wissenschaftliche Medizin*, if the piece could be given a tone sufficiently "physiological" to satisfy its editor, Emil Du Bois-Reymond. If his work involved histology, he might well send it to Jakob Henle's *Zeitschrift für rationelle Medizin*, which ran from 1844 to 1868. After 1865 he had the option of Max Schultze's *Archiv für mikroskopische Anatomie*, which also published essays on cell development, single-celled organisms, and new techniques of preparing materials for

10. The *Archiv für Naturgeschichte*, originally published under the editorship of A.F.A. Wiegmann, was founded in 1835 and continued publication until 1926. The language of the *Zeitschrift*'s prospectus so closely parallels that of the *Archiv*'s prospectus, written in the 1830s, that one suspects the latter was used as a model for the former. The editors of the *Zeitschrift* merely inverted the language of the *Archiv*'s prospectus to describe their own emphasis. Compare the passage quoted above with this one from the *Archiv*: "Essays in descriptive zoology, descriptions of new genera and species, reports on the mental abilities, mode of life and geographical distribution of already-known animal species, even anatomical [*zootomische*] reports, insofar as these justify or secure the systematic position of an animal or a whole group, will find a suitable place here." "Prospectus," *Archiv für Naturgeschichte* 1 (1835): n.p.

11. Siebold to Ehlers, 6 May 1874 (no. 70). Universitätsbibliothek Göttingen, Cod. Ms. E. Ehlers 1815. Hereafter all files in the Ehlers Nachlass (Cod. Ms. E. Ehlers) will be designated by "EE" followed by the file number, as in EE 1815.

microscopical study.[12] If the essay concerned the life cycle of a human parasite, he might even try a journal of clinical medicine or pathology.

Finally, a very likely place for a zoologist to publish his early work was not a zoology journal at all, but the proceedings or journal of the local scientific society at or near his university, where he might deliver an oral version before submitting it for printing in the group's journal or proceedings. These societies, such as Würzburg's *physikalisch-medizinische Gesellschaft* and Jena's *medizinisch-naturwissenschaftliche Gesellschaft*, were scientific groups of wide-ranging interests, and their journals served mainly to foster local ties rather than disciplinary ones.[13]

Among these various journals the *Zeitschrift für wissenschaftliche Zoologie* stood out as one of only two specifically devoted to zoology, and the only one to conceive zoology in terms broader than systematics. Although in its early years, especially, it drew contributions from physiologists, anatomists, and even a few botanists, it was a professional journal in the sense of being directed toward one special realm of scientific inquiry, rather than seeking a broader audience among scientifically inclined people. As zoology gradually attained distinct disciplinary status throughout the German universities, the *Zeitschrift* was there to provide practitioners of "scientific" zoology with an outlet for technical communication.

What these zoologists were communicating to each other was thoroughly grounded in the empirical side of zoology. Even after Kölliker effectively withdrew from editing in the late 1860s and Ernst Ehlers began managing its day-to-day business in 1874,[14] the *Zeitschrift* remained true to

12. Otto Bütschli, a frequent contributor to the *Zeitschrift*, also published in the *Archiv für Naturgeschichte* and Du Bois-Reymond's and Schultze's journals in the early and mid-1870s. For a list of his publications, see A. Kossel, *Das Lebenswerk Otto Bütschlis* (Heidelberg: Carl Winter, 1920), also printed in *Sitzungberichte der Heidelberger Akademie der Wissenschaften*, Mathematisch-naturwissenschaftliche Klasse, Abt. B. Biologische Wissenschaften. Jahrgang 1920. 1. Abhandlung. August Weismann, about whom more below, published a number of articles in Henle's journal and two in the *Archiv für Anatomie, Physiologie und wissenschaftliche Medizin* in the late 1850s and early 1860s.

13. See Albert von Kölliker, "Festrede zur Feier des 25 jährigen Bestehens der Physikalisch-medizinischen Gesellschaft in Würzburg," *Verhandlungen der Würzburger physikalisch-medizinische Gesellschaft* N.F. 9 (1874). On the distinction and tension between local and disciplinary allegiances in Germany, see R. Steven Turner, "The Growth of Professorial Research in Prussia, 1818 to 1848—Causes and Context," *Historical Studies in the Physical Sciences* 3 (1971): 137–182. Jungnickel and McCormmach also use these two contexts as the chief ones structuring the social aspects of their analysis of the development of theoretical physics in Germany. (*Intellectual Mastery of Nature*, xviii.)

14. In 1874 Siebold noted that by his count, between 1867 and 1883 he himself had sent 110 manuscripts to the publisher for printing, whereas Kölliker had submitted only 27. Siebold to Ehlers, 6 May 1874 (no. 70). In 1877, Siebold again referred to Kölliker's lack of editorial

the aims set forth in 1848. Under Siebold's enduring guidance it published a steady stream of "hard," factual scientific research in comparative anatomy, histology, embryology, and physiological morphology, by an increasingly broad circle of authors. Naturally, new topics came to the fore, new theoretical problems were treated, argued over, and sometimes put to rest. Beginning in the late 1860s, for example, cautious discussions of Darwinism began to appear; a few years later the debate over the germ layer theory found its way into the pages of the *Zeitschrift*. Yet the overall tone of the journal remained remarkably consistent. Theory was important, to be sure, but in developing theory facts came first, and wherever theoretical issues were raised the authors sought to couch their arguments in the most "objective," fact-based terms.

The articles themselves presented only the public face of the enterprise, however, and the unpublished side was just as important. Siebold envisioned his journal as a community project. Authors who frequently published in his journal were not simply contributors, but *Mitarbeiter* (literally, "coworkers"), a word suggesting that all are contributing to a common enterprise. While the term can also be translated more neutrally as "colleagues," in Siebold's frequent uses of it he clearly meant something more: he meant regular contributors involved in an ongoing community enterprise. The decision to accept or reject a work, therefore, involved not only deciding its value as a piece of scientific research, but also judging the contributor himself.

To have one's first piece accepted by the *Zeitschrift* was to gain entry into the ranks as a potential *Mitarbeiter*, and given the significance of the editors' decision, the paper's merits alone would not suffice to have it printed. One had to be known to one of the editors, or to be recommended to them by a *Mitarbeiter* as a good scientific worker. In a typical case, the University of Vienna's professor of zoology Carl Claus, who had close ties to Siebold, sent a paper by a student of his named Richter to Siebold for publication in the *Zeitschrift*. Siebold wrote Claus that although he had not had a chance to read it carefully, he was willing to print it if he could be assured that it was sent with Claus's recommendations and that the research had been conducted under his supervision. A few years later another letter from

participation: "currently he has no taste and no disposition for [even] the smooth editorial business." Siebold to Ehlers, 23 February 1877 (no. 95), EE 1815. Kölliker's name remained on the masthead until he died in 1905, but Siebold was really in charge by the mid-1870s, and Ehlers gradually took over. Upon Siebold's death in 1885, Ehlers was named co-editor with Kölliker, and he remained editor until his own death in 1926.

Siebold queried Claus about his opinion of Berthold Hatschek, a name new to Siebold. He had received a paper from Hatschek for submission to the *Zeitschrift*, the research for which had been conducted at the marine station Claus ran in Trieste. The paper did not mention Claus's role in the work, however, and Siebold could not recall Claus's ever mentioning Hatschek's name. Siebold was therefore writing to ask Claus his opinion of Hatschek's work: "Before I know something about the *curriculum vitae* of a new *Mitarbeiter* for the *Zeitschrift f. w. Z.*, I don't like to decide on taking his work."[15] This revealing sentence reinforces the impression that the quality of the work itself was only a part of the decision to accept it or not. The commitment would not be just to accepting this particular piece of work, it seems, but to bringing the writer into the club.

Once taken into the fold, an author could expect to have his work accepted without further ado, as long as he followed the general editorial policy of contributing original research. Under Siebold's editorship, this commitment extended so far that certain *Mitarbeiter* did not even have to send him their works for editorial approval—they could send them directly to the publisher for printing in the journal. Until the editors changed their policies in the late 1870s, those zoologists especially close to Siebold, such as Claus, Rudolf Leuckart, and Ernst Haeckel, not only had *carte blanche* for their own work, but could also submit works of others with their recommendations straight to the publisher.[16]

This casual approach to manuscript submissions meant that the *Zeitschrift*'s publisher, Wilhelm Engelmann, also had license to exercise his editorial judgment about a manuscript that was sent directly to him, sometimes forwarding it to the editors for approval even when the author had informed him the piece could be published without further editing. Since Engelmann also published other serials and books relating to zoology, he was quite conversant with what was going on in the field, and he had inside knowledge that was sometimes useful to the editors. For example, in late 1876 Engelmann wrote Ehlers that he had received the brief summary of research by a certain Dr. B. Gabriel in Breslau that Ehlers had forwarded for

15. Siebold to Claus, 27 July 1873, (Bl. 70); 7 July 1877 (Bl. 89), Berlin, Staatsbibliothek Preussischer Kulturbesitz, Nachlass Carl Theodor Ernst von Siebold, 3k, 1855(5). Hereafter BSPK 3k 1855(5).

16. Siebold to Ehlers, 26 April 1874 (no. 69), EE 1815. As early as 1865 Siebold had written Claus, "your works will naturally find an open, enthusiastic inclusion in the *Zeitschrift für wissenschaftliche Zoologie* at any time." 25 May 1865 (Bl. 52), BSPK 3k 1855(5). It should be noted that although Siebold included Haeckel on his list, the latter had stopped contributing to the journal in 1869.

publication in the *Zeitschrift*. Engelmann thought Ehlers should know that another journal's editor had already rejected this work two or three times, and that a previous essay had been accepted for publication in that journal only through the direct intervention of the anatomy professor at Breslau, Carl Hasse. As a favor to Hasse, Engelmann had also printed the same earlier essay *gratis* as Gabriel's doctoral dissertation. He warned Ehlers, "If he is accepted by our journal, he will write bundles-full, and you will never again be free of him, and [he will be] a great danger to our journal. I consider it my duty to tell you all this, because there is still time to keep away from him."[17] When Siebold was consulted by Ehlers, he agreed that a rejection was in order, writing that "we lose nothing, if this *Mitarbeiter* escapes our journal; in any case, with the endorsement that he enjoys, his résumé will find accommodation somewhere."[18]

A similar case arose a few months later when Karl Hennig, professor extraordinary of obstetrics at Leipzig, submitted a manuscript in person to the publishing house, saying it was not necessary to send it to the editors for a look before publishing it. Wilhelm Engelmann's son Rudolf, who had just entered the business,[19] accepted the manuscript and tentatively marked it in for volume 29. When he sent Siebold a list of upcoming articles, the latter requested the manuscript be sent to him, because he did not know Hennig well enough to accept it sight unseen. Upon reading it, he was dismayed to discover how poorly it was written and that it was evidently the continuation of another work published somewhere else. Siebold wrote Ehlers that it could not be accepted in its current form; at the least it needed a suitable introduction and conclusion. Siebold agreed with Ehlers's eventual decision to reject the piece and again couched his opinion not just in terms of the piece, but of its author: "That you have enticed Herr Hennig away from being a *Mitarbeiter* of our journal pleases me. We really have no lack of participants; indeed, we are always so far ahead of almost all the other rival journals, that we can even reject some things, especially something so mediocre as that concoction of Hennig's."[20]

17. Engelmann to Ehlers, 18 Nov. 1876, enclosed in Siebold to Ehlers, n.d. (no. 87), EE 1815. Because dissertations had to be published, it was common for dissertators to use an article simultaneously as their dissertation, requesting a new title page from the publisher along with the number of copies required by the university. (By the mid-1850s scientific dissertations were generally published in German rather than Latin.) In the normal course of events, the dissertator paid for the extra copies himself, but because the essay was already typeset except for the title page, it was considerably cheaper than printing the dissertation entirely separately.

18. Siebold to Ehlers, n.d. (no. 87), EE 1815. The endorsement referred to is presumably that of Hasse.

19. *Jubiläumskatalog der Verlagsbuchhandlung Wilhelm Engelmann in Leipzig, 1811–1911* (Leipzig: Druck von Breitkopf u. Härtel, 1911), 94.

20. Siebold to Ehlers, 13 January 1877 (no. 92); 4 March 1877 (no. 97), EE 1815.

From these examples it should be evident that both the acceptance and rejection of manuscript submissions were often explicitly (if privately) tied to the presumed character and overall scientific reputation of the articles' authors. The reason for this should by now be clear: because accepting a piece of work meant opening the circle to a new member, the author—not just the article—had to pass judgment for acceptance. Personal recommendations (or warnings) as to the writer's character and promise were therefore a critical part of the process.

Satisfactory personal characteristics or recommendations from *Mitarbeiter* weren't always enough to guarantee publication, however; it was still sometimes possible to reject a work if it failed to meet the editors' criteria for a "substantial" contribution to science. From Siebold's letters it is evident that a lack of substance could come in many forms. First, the article might not be sufficiently original. As early as 1854, Siebold rejected a piece (on the avian digestive tract) recommended by Julius Victor Carus, professor of comparative anatomy at Leipzig. Siebold admitted it was a nice piece of work, but said that it merely reconfirmed well-known results, an uninteresting outcome that "is to be forgiven in a dissertation" but "a waste for a journal."[21] To take another example, in 1874 he received a piece by a young zoologist named Dewitz, along with the recommendation of Dewitz's teacher, the zoology professor Zaddach at Königsberg. The paper, on the structure and development of the stinging apparatus in ants, appeared to Siebold to cover just the same territory as one that had appeared the previous year, a paper which, moreover, had won a prize. The work by Dewitz could be accepted only if, as Zaddach claimed, it really was substantially better than the earlier paper. Siebold put the decision in Ehlers's hands, claiming that he was too busy with his own research to read the Dewitz paper more closely. Ehlers evidently decided it was indeed good enough to publish (perhaps with some revisions), for it appeared three years (!) later in volume 28 of the *Zeitschrift*.[22]

Alternatively, an article might be rejected for conveying results that were too preliminary to be adequately substantiated. Although the *Zeitschrift* occasionally published brief preliminary reports of research, most papers were at least ten pages long (in rare cases they ran as long as a couple of hundred pages), and a preliminary report was rarely accepted as a *Mitarbei-*

21. Siebold to J. V. Carus, 7 Oktober 1854 (Bl. 33), BSPK 1855(5).
22. Siebold to Ehlers, 10 June 1874 (no. 71), EE 1815. H. Dewitz, "Ueber Bau und Entwicklung des Stachels der Ameisen," *ZwZ* 28 (1877): 174–200. The earlier, prize-winning paper was Carl Kraepelin, "Untersuchungen über den Bau, Mechanismus und die Entwicklungsgeschichte des Stachels der bienenartigen Thiere," *ZwZ* 23 (1873): 289–331.

ter's first contribution to the *Zeitschrift*. In response to Ehlers's suggestion in late 1874 that they might expand their operation to publish a newsletter with a shorter publication lag time, to accommodate many more preliminary reports, Siebold expressed his doubts as to the benefits. He wondered whether such a publication would not encourage

> a quantity of raw research to be sent out as *preliminary* reports, in order to ensure *preliminarily* the right of priority. Should the subsequent, more leisurely repeated investigations produce no confirmation, then one must leave the precipitately reported material go until it has lost itself in the sand of oblivion, and be happy if no one speaks of it.[23]

Such works did not advance the cause of science, but only wasted time and money. It was better, in Siebold's view, to wait for the mature, finished product.

For similar reasons articles primarily of a speculative nature were also unwelcome in the *Zeitschrift*. Referring to an article submitted in 1875 by a *Privatdocent* named Hermann von Ihering, Siebold wrote,

> Let us leave such far-reaching speculative utterances . . . for other journals. . . . We want to reserve the space in our journal for facts. The more new facts our journal brings, the more it will be necessary for the speculative spirits of our time to change their principles again and again, whereas that which is fact will stand.[24]

Finally, the one thing Siebold absolutely would not tolerate was polemic, which was to be shunned as an effort that did not contribute anything of substance to science and as conduct unbecoming a gentleman and a scientist. As Siebold put it to Ehlers in late 1873, "You guessed correctly when you attribute to me an aversion to accepting polemical expectorations in the *Zeitschrift*. . . . Up to now our journal has managed to stay free of such quarrels." In a later letter in which he laid out his general editorial principles to Ehlers, he said that although he never changed the text so as to interfere with its sense, "rude remarks toward other authors, if I were to catch sight of them (which however has never yet happened), I would not tolerate."[25] All too soon Siebold would have to act upon these beliefs.

Siebold's approach to editing was based heavily on trust and personal

23. Siebold to Ehlers, 15 December 1874, (no. 75), EE 1815.
24. Siebold to Ehlers, 23 December 1875, (no. 84), EE 1815.
25. Siebold to Ehlers, 17 October 1873 (no. 62); 6 May 1874 (no. 70), EE 1815.

cooperation. He trusted some of his coworkers to the extent that they did not even have to show him their work before he published it. He relied on his colleagues to judge their students' merits, accepting their word that the work was competently done. And he assumed that once in the fold, his *Mitarbeiter* would continue to abide by the canons of publishing behavior that they had had to display to get into the club in the first place. They were supposed to continue the dogged empirical work of the true scientist; they were *not* supposed to betray the trust their editor had placed in them by asking him to print the sorts of writings that he would automatically reject in a newcomer. In the late 1870s, however, Siebold and Ehlers had to confront just such a breach of trust when two longtime contributors to the *Zeitschrift* became embroiled in a bitter dispute over water fleas.

The Lesser Daphnoidian Controversy

August Weismann (1834–1914) and Carl Claus (1835–1899) counted among the leading lights of the younger generation of zoology professors in the 1870s. Only a year apart in age, they had similar scientific backgrounds. Both had followed the most common route of students interested in biology, taking degrees in medicine while pursuing their biological interests. Both found Darwin's theory attractive and from the late 1860s published research and review essays in support and elaboration of it. Perhaps most important, both studied under Rudolf Leuckart (1822–1898), the leading zoological teacher of his generation. Under Leuckart's influence Weismann and Claus learned to study the organism not just to determine its taxonomic locus, but also to elucidate the relationship between its form and function. As for most zoologists of their time, the study of the organism's life cycle and its relationship to its environment was an especially important aspect of this pursuit. Given the degree to which these interests coincided with the aims of the *Zeitschrift*'s editors, it is not surprising that both began publishing there early in their careers (Claus in 1860, Weismann in 1863). By the early 1870s they counted among the journal's most steadfast contributors.

In the early 1870s both men began studying organisms that fall into the crustacean order *Cladocera* or, to use the older name preferred by Weismann, *Daphnoidea*. These creatures, commonly known as water fleas, were especially amenable to study because their bodies were transparent throughout development, and one could follow the morphological changes in the live animal by placing it on an indented slide directly under the microscope. Although one of the two families within this order, the daphnids, had been carefully studied by Franz Leydig in the 1850s,[26] the possibilities of the order had certainly not been exhausted. The other family, the polyphemids, had received considerably less attention by German workers, and even within the daphnids the natural history of certain genera were less than fully worked out. It was to these areas that Claus and Weismann turned in the mid-1870s.

Their dispute in the *Zeitschrift* began in January 1877, with the introduction to the second part of Weismann's four-part article "On the natural history of the Daphnoids." Just when he had sent the manuscript of Parts II through IV to the publisher, Weismann wrote, he was shocked to learn that Claus had just published a "preliminary report" of researches into the polyphemids, the very group he himself had been studying. Worse yet, Claus had discovered the very same phenomenon that formed the centerpiece of Weismann's Part III—the fact that the eggs were not self-contained sources of nutrition for the embryo, but received nutriment directly from the "brood chamber" surrounding them inside the mother. Weismann accused Claus of having followed him into this territory and complained that while he had kept Claus informed "through friendly letters" of what he was working on, Claus had left him in the dark on his own plans and progress. He finished the introduction by charging that Claus, knowing full well what Weismann was working on, had rushed to extend his own researches into the same area to beat him into publication.[27]

Two numbers later, in the issue dated 23 April 1877, Claus published a four-page response, "In correction and defense," in which he argued that he certainly had not followed Weismann into this territory, that they weren't even studying the same thing. Weismann was interested only in the development of eggs. He, Claus, was interested in a broader (and, implicitly, more important) problem: elucidating the anatomy and development of the overall organism. Given this aim, it would hardly be appropriate for him to leave out a section on egg development. He reprinted excerpts from letters Weismann had sent him between 1874 and 1876 to show readers just how vague Weismann's claims to priority were. He also asked whether "Herr Weismann believes that through these epistolary remarks he had ensured a claim to the research field of the daphnoids and their egg-

26. See F. Leydig, *Naturgeschichte der Daphniden* (Tübingen, 1860).
27. August Weismann, "Beiträge zur Naturgeschichte der Daphnoiden, Theil II, III u. IV," *ZwZ* 28 (1) (30 January 1877): 93–254. Introduction (93–95), dated 23 November 1876.

development?" Taking the high ground, he argued that the simultaneous, *independent* appearance of works on the same subject was good for science. He did not remain there for long, however, but concluded by referring to the "insinuating remarks" Weismann had made "purely to pick a quarrel," insinuations that were entirely a "fiction" of their author.[28]

Weismann's response appeared in the *Zeitschrift* in November. In it he repeated his main claims, stating, however, that he had been in no way staking an exclusive claim to research territory, which would have been unscientific, but that nevertheless he found Claus's behavior "not very friendly, in a manner of speaking disloyal." In showing how Claus had rushed into publication, he pointed to errors in Claus's observations and argumentation that would have been avoided with more careful work. After numerous examples, he finished impugning Claus's thoroughness and honesty by claiming that Claus had cited a major work (published, to be sure, in Danish) without reading it. He concluded the piece with a curious four-paragraph discussion on the language problem in science, arguing that scientists should be forced to write in German, English, or French so that no one would have to slog through a work written in Danish or some other minor language.[29] To the end of this piece the editors appended a note stating: "With this 'Defense' the editors of the journal view this dispute, as far as it concerns personal relations, as closed."

What were the editors thinking of, in allowing such a nasty dispute to sully the pages of their journal? As far as Siebold could remember, they had never allowed such vituperations to come to print before.[30] Given the strong feelings he had expressed in rejecting similar invective just a few years earlier, why did he feel he had to accept it *now*? It certainly was not the case that he had changed his mind about polemics. When he first received Claus's response to Weismann's original priority claim, Siebold wrote a polite and apologetic letter rejecting the essay.

> I cannot resolve to allow such a controversy as the printing of this response would spark, and which could only be uncomfortable to the readers of this journal. A priority dispute, which mostly interests the disputing persons more than the reader who stands outside of it, always has something unpleasant about it. . . . You can surely understand how unpleasant such a step [as printing Claus's response] would be for me, which would have the conse-

28. Carl Claus, "Zur Berichtigung und Abwehr," *ZwZ* 28 (4) (23 April 1877): 571–574.
29. August Weismann, "Rechtfertigung," *ZwZ* 30 (1) (30 November 1877): 194–202.
30. Siebold to Ehlers, 23 February 1877 (no. 95), EE 1815.

quence that two *Mitarbeiter* whom I value so highly, and who for years have bestowed upon my journal their splendid brain-children, should antagonistically fall out.[31]

Yet he did print Claus's response, and Weismann's rebuttal as well. Why? The answer appears to be entwined with Siebold's concept of the *Mitarbeiter* and the fealty he owed them. Having once rejected Claus's heated initial response, Siebold found it difficult to reject his "mitigated (as he thinks)"[32] resubmission. He would have been sorry to see Claus go off in a huff with his rejection, but it clearly pained him that the latter refused to call the matter closed, especially since in Siebold's own view Claus had brought the initial accusation upon himself. Since Claus refused to drop the issue, Siebold felt compelled to accept a version of his response, but even his "milder" resubmission seemed too harsh to print. This raised real problems for Siebold, who, it will be recalled, disliked tampering with his authors' prose. Worse yet, he worried that the pugnacious tone of Claus's response might well occasion a rebuttal from Weismann, which would only extend the unpleasantness. Siebold agonized over what to do in letters to Ehlers and Kölliker, and eventually the three agreed to ask Claus to withdraw the second, more offensive half of his response, to which request he acceded. Even after this, though, Claus could not resist inserting words into the page proofs which Siebold felt obliged to remove.[33]

Siebold had some hopes that the final version of Claus's response would not elicit a response from Weismann, but those were not fulfilled. Weismann's rebuttal, too, was returned for toning down, and when he resubmitted it, he said that he had mitigated it as far as it was possible. What remained, he said, was only that which he could *prove*; true to the form of the *Zeitschrift,* he claimed to "let the facts speak for themselves." Furthermore, he said, this was not just a private, personal matter, but a question of "a general principle, . . . the *principle of decorum in science!* For if a Claus allows himself to use such unclean means to cheat on others, we run the risk that not only he will acquire a taste for this sort of robber-knightdom, but that in the end he will found an entire school of robber-knights!"[34]

31. Siebold to Claus, 20 Feb. 1877 (Bl. 80), BSPK 3k 1855(5). Over twenty years earlier, Siebold had expressed similar discomfort with priority disputes in a letter to his friend Rudolf Wagner. Siebold to Wagner, 29 October 1854, (no. 23). Universitätsbibliothek Göttingen, Cod. Ms. R. Wagner 7.
32. Siebold to Ehlers, 23 February 1877 (no. 95), EE 1815.
33. Siebold to Ehlers, 23 February 1877 (no. 95), 28 February 1877 (no. 96), 4 March 1877 (no. 97), 12 March 1877 (no. 98), 7 April 1877 (no. 68), EE 1815. Siebold to Claus, 26 March 1877 (Bl. 84), BSPK 3k 1855(5).
34. Weismann to Siebold, 13 August 1877, EE 2092 (Weismann).

Siebold forwarded Wiesmann's revised rebuttal to Ehlers with the comment, "I believe we must do our excellent *Mitarbeiter* Weismann the favor of taking this essay for the *Zeitschrift*."[35] At the same time, he found himself very uncomfortable with the entire business, perhaps most of all because, he said, "I fear that with this controversy we might lose one of the two as long-term *Mitarbeiter*, which I would very much regret." Again later he wrote Ehlers that he feared Claus would "withdraw from the journal, which would be very regrettable."[36] Siebold worked hard to retain both his valued contributors: to keep Claus happy he compromised his principles on polemics, and then to satisfy Weismann he had to extend the controversy. The ultimate result left no one happy, least of all Siebold, who remained with the knowledge that in his desire to placate his contributors he had allowed them to rupture the communal spirit of his journal.

Although the correspondence surrounding this affair is incomplete, the evidence suggests that Weismann was more successful than Claus in appealing to his editors. He presented himself as a good *Mitarbeiter*: his published case (he claimed) rested on "the facts," and in his unpublished case he presented himself as serving the cause of scientific decorum, which was certainly part of Siebold's view of a good *Mitarbeiter*. By contrast, in both his published and unpublished remarks he presented Claus as violating the bounds of proper scientific behavior. A letter to Ehlers illustrates this twofold presentation well. After his resubmission was accepted, he asked to take it back still again to make it even milder—surely the step of a gentleman who believed his behavior was uncomfortably indecorous. This request was accompanied, however, by a threat that if Claus were allowed another response, Weismann would be forced to get nasty and tell the entire sordid story, which as yet was untold. He claimed, in fact, to have proof that Claus had changed his findings at the page proof stage after reading Weismann's essay and had then back-dated his own essay to claim priority.[37] Once again, a concrete, "fact-based" accusation served to malign Claus's reputation and honesty.

Claus, by contrast, was hotheaded and unrepentant. He would not tone down his remarks to the satisfaction of the editors, causing them to take the uncomfortable step of altering his prose. He refused to let the matter rest when the editors declared in print that it was finished, but insisted they print his counterrebuttal to Weismann's rebuttal, saying he "regretted not being able to spare the editors this step." (Note his assumption that they

35. Siebold to Ehlers, 19 August 1877 (no. 100), EE 1815.
36. Siebold to Ehlers, 23 February 1877 (no. 95), 12 March 1877 (no. 98), EE 1815.
37. Weismann to Ehlers, 4 September 1877, EE 2092.

had no choice in the matter.) He further chastised Ehlers for not having read his article carefully enough to have known that Weismann's attack was unfounded; had Ehlers done so, he would naturally have refused to print it in the first place. When Ehlers pointed out that it was not his job to look out for Claus's interests, Claus responded with a long, almost incoherently angry letter which argued, among other things, that "it is not a question of your intervening for *my person*, but of *the objectivity of the editorship*."[38] When Ehlers returned his counterrebuttal with the editors' collective decision not to print it, Claus took it to the local zoological-botanical society in Vienna and had it printed in its proceedings. In his ten pages of remarks not only Weismann's honesty but also the wisdom of the *Zeitschrift*'s editors came under his fire.[39] All in all, Claus did *not* act as a good *Mitarbeiter* should, and it is perhaps not surprising that after this affair he never again published in the *Zeitschrift*.

At the height of the controversy Siebold remarked, "The young *Herr* zoologists seem to be frightfully sensitive, and if things go on like this it will spoil the editing of a journal."[40] For Siebold, this was certainly true, for in his view of things editing the journal went hand in hand with managing the cooperative venture that was zoological research. Editing the journal could be a pleasure or, indeed, a task that could be carried out at all, only as long as an atmosphere of trust and cooperation could be maintained, for without it the entire system upon which the acceptance of papers was based—the very community that provided the journal's *raison d'être*—threatened to collapse. And yet that was just what was happening.

Publishing Polemics

It would be tempting to dismiss the dispute between Claus and Weismann as a fluke, the result of a personality clash between two ill-tempered, oversensitive scientists. The fact that it appeared in print (one version of it, anyway) showed how far it was possible to push one's privileges as *Mitarbeiter* if one wanted to. But this was not an isolated incident. In the late

38. Claus to Ehlers, 14 December 1877 (no. 2); Ehlers to Claus, 22 December 1877 (transcript in Ehlers' hand; Beilage 1); Claus to Ehlers, 24 December, 1877 (no. 3), EE 279.
39. Ehlers to Claus, 6 January 1878 (Beilage 2), EE 279. Carl Claus, "Anlass und Entstehung seiner eigenen Untersuchungen auf dem Daphniden-Gebiete," *Verhandlungen* der k.k. zoologisch-botanischen Gesellschaft in Wien, Versammlung am 6 Februar 1878 [1879], 28: 6–16.
40. Siebold to Ehlers, 28 February 1877 (no. 96), EE 1815.

1870s and early 1880s conflicts over priority were suddenly cropping up all over the zoological community. Claus seems to have been involved in a number of them, as Weismann was quick to point out to Siebold. Weismann himself was not immune to other such conflicts, however, and in fact engaged in priority disputes with the very same two zoologists he mentioned as fighting with Claus, namely Otto Bütschli and Alexander Goette.[41] Others were engaging in similar disputes as well. In one of the least savory cases, a doctoral candidate accused a *Privatdocent* (lecturer) who had been his teacher of plagiarizing from his unpublished dissertation on the nervous system of nematodes while it was circulating among the faculty. When the *Privatdocent* denied the charges the student escalated his attack, claiming that the former had never worked on nematodes and had known *nothing* about the peripheral nervous system of the nematode *Ascaris megalocephala* until the latter had told him informally of his own results. The *Privatdocent* countered that he had invited the student to study *his own,* technically more proficient, slides while he was out of town. The published record of the dispute ends there, but it should be noted that the student, Emil Rohde, eventually became a professor, although little more was heard from the *Privatdocent.*[42]

A striking aspect of all these unpleasant events is how many of them saw print in journals or other serial publications. Earlier, the most common way of publishing a polemic was to print an independent pamphlet—a genre that placed the writer in a position of autonomy relative to his community, but in which he also risked being ignored. Although publishing pamphlets remained one ready way to vent one's spleen, beginning in the 1870s new outlets became available for polemical writings. Among the many new journals launched in this period that published the writings of zoologists, two kinds especially served to fan the polemical flames: the institute-based serials and those that published short summaries of research results.

In 1872 the professor Carl Semper began publication of a series called

41. Weismann to Siebold, 13 August 1877, in EE 2092; Weismann to J.V. Carus, 20 July 1880 (38) and 19 May 1882 (46), BSPK Lc 1889, August Weismann.

42. See Emil Rohde, "Einige Erklärungen zu 'Vorläufige Bemerkungen über Muskulatur, Excretionsorgane und peripherisches Nervensystem von Ascaris megalocephala und lumbricoides von Dr. Gustav Joseph' in No. 125 des *Zoologischen Anzeigers*," *Zool. Anz.* 6 (131) (5 February 1883): 71–76. Gustav Joseph, "Erwiederung auf die Erklärungen des Herrn Dr. Rohde im *Zoologischen Anzeiger* No. 131 (5 February 1883)." *Zool. Anz.* 6 (133) (5 March 1883): 125–127. Emil Rohde, "Über die Nematodenstudien des Herrn Dr. Joseph," *Zool. Anz.* 6 (136) (16 April 1883): 196–199. Gustav Joseph, "Zur Abwehr gegen die ferneren Angriffe des Herrn Dr. Rohde auf p. 196–199 des Zoolog. Anzeigers: 'Über die Nematodenstudien' etc." *Zool. Anz.* 6 (139) (21 May 1883): 274–278.

Arbeiten aus dem zoologisch-zootomisch Institut zu Würzburg. Anton Dohrn, head of the new Zoological Station at Naples, and Carl Claus in Vienna followed suit in 1878, and by 1886 the heads of six zoological institutions were putting out their own periodical volumes of research results.[43] As well as providing a ready vehicle of publication for their students and assistants, these serials allowed their editors to sound off without fear of censorship. Semper took full advantage of his editorial autonomy, publishing a series of biting critiques of other zoologists' works that would have stood small chance of finding a home in a research journal edited elsewhere. He also allowed researchers at his institute to advance their quarrels in its pages: Robby Kossmann, who had published his *Habilitationsschrift* in Semper's journal, later used the journal to defend himself against Anton Dohrn's intimation that Kossmann had appropriated results and ideas from him as his own. In his defense, Kossmann also took time to attack Dohrn's work as speculative and not based on serious research. This little polemic was accompanied by a note from Semper, saying he welcomed it "with the greatest pleasure" because Kossmann so clearly identified and denounced Dohrn's attempt to "introduce a new principle in science: taking half-yellowed pages from the wastebasket to justify claims to property."[44]

Claus, surprisingly, rarely used his journal to blast his opponents. He did devote an essay in 1882 to attacking Ernst Haeckel's recent monograph on the medusae, beginning with the accusation that Haeckel had copied Claus's own findings without acknowledgement and had merely created a new nomenclature to present them as original; Haeckel had also misconstrued Claus's results in order to dismiss them.[45] But Claus spewed most of his invective elsewhere, publishing pamphlets, firing off heated notes to editors of other journals in protest against other scientists' abuse of him (mostly with regard to priority), and even inserting a number of

43. *Arbeiten aus dem zoologisch-zootomischen Institut zu Würzburg* (1872–1895), ed. Carl Semper; *Mittheilungen aus dem königlichen zoologischen Museum zu Dresden* (1875–, with various name changes),ed. A.B. Meyer; *Mittheilungen aus der zoologischen Station zu Neapel,* (1878/9–1902), ed. Anton Dohrn; *Arbeiten aus dem zoologischen Institut der Universität Wien und der zoologischen Station in Triest* (1878–1915), ed. Carl Claus; *Zoologische Beiträge* (works out of Anton Schneider's zoology institute in Breslau) (1883–1893), ed. Anton Schneider; *Arbeiten aus dem zoologsichen Institut zu Graz* (1886–1911), ed. Ludwig Graff.

44. Anton Dohrn, *Der Ursprung der Wirbelthiere und das Princip des Functionswechsels* (Leipzig: Engelmann, 1875). R. Kossmann, "Die Ansprüche des Herrn Dr. Dohrn auf Lösung des Rhizocephalen-Problems," *Arbeiten aus dem zoologish-zootomisch Institut in Würzburg* 2 (1875): 510–515.

45. Carl Claus, "Zur Wahrung der Ergebnisse meiner Untersuchungen über Charybdea als Abwehr gegen den Haeckelismus," *Arbeiten aus dem zoologischen Institut der Universität Wien und der zoologischen Station in Triest* 4 (1882): 1–14. (Each contribution is separately paginated.)

strongly flavored comments about others into later editions of his zoology textbook.[46]

Haeckel, one of Claus's chief targets, contributed his share and more to the contentious atmosphere. Although he did not have an institute-based journal *per se*, the *Jenaische Zeitschrift für Medizin und Naturwissenschaft* became virtually a house organ for him and his followers. Haeckel was the self-proclaimed apostle of Darwinism in Germany—we might call him "Darwin's schnauzer"[47]—and he took his mission seriously. One of the great polemicists of his day, Haeckel had a way of tying up his own far-reaching evolutionary speculations with dismissive comments about almost all other research previously conducted in any of the areas in which he ranged. For example, in an essay published in the *Jenaische Zietschrift* in 1877, he reviewed the reception of his "gastraea theory," which sought to provide the empirical and theoretical basis for believing in evolutionary recapitulation. In the four years since he had first presented it, he wrote, many well-known zoologists (he listed six) had come up with results that supported the theory, and the most important objections to it had been resolved. (He does not say what these were.) "For this reason it . . . appears superfluous to answer the fierce attacks that were immediately directed at the gastraea theory and its consequences by Carl Claus, Carl Semper, W. Salensky, Alexander Agassiz and others."[48] Haeckel's speculative approach to science, combined with his cavalier treatment of other scientists' ideas, made him a lightning rod for polemical attacks from his colleagues.[49]

46. Claus, *Die Typenlehre und E. Haeckel's sogenannte Gastraea-Theorie* (Vienna: Manz'schen Buchhandlung, 1874); "Erklärung in Betreff der Prioritätsreclame des Herrn Ed. van Beneden," *Zoologischer Anzeiger* 3 (50), (8 March 1880): 107–110; "Zur Prioritätsreclamation des Herrn Dr. Y. Delage," *Zoologischer Anzeiger* 8 (197) (15 June 1885): 356–357. In his *Lehrbuch der Zoologie*, 4th ed. (Marburg: Elwert, 1887), for example, he speaks sharply of Moritz Wagner and Weismann (181), Haeckel (102–103), and Carl Nägeli (186–189).

47. Since much of this essay is about credit, it seems proper to note that John Beatty and I have independently been using this appellation in talks and lectures for a number of years, although we only recently discovered the coincidence. Haeckel's position in Germany was not perfectly parallel to that of T. H. Huxley, "Darwin's bulldog," in England, but it was similar enough to make us look for an analogy. Anyone who has seen a picture of Haeckel will understand why we reached for this particular label; it is so much more appropriate than, say, "Darwin's dachshund."

48. Ernst Haeckel, "Nachträge zur Gastraea-Theorie," *Jenaische Zietschrift für Naturwissenschaft* 11 (1877) (N.F.4): 55–98, on p. 55.

49. In addition to the essay by Claus discussed briefly above, see his *Die Typenlehre und Haeckel's sogenannte Gastraeatheorie* (Vienna, 1874). Other attacks by zoologists include Carl Semper, *Offener Brief an Herrn Prof. Haeckel in Jena* (Hamburg: W. Mauke Söhne, 1877); Robby Kossmann, *Eine Warnung gegen E. Haeckel's Citaten* (Heidelberg: Winter, 1897); and J. W. Spengel, "Hyper-Darwinismus und Anti-Darwinismus," *Gaea* 10 (1874): 329–334. A more complete list of titles of responses to Haeckel up to 1878, many of them with a critical cast, may be gleaned from O. Taschenberg, *Bibliotheka Zoologica*, 2d series 1.

The polemical exchanges between Haeckel and others also served as a model for the many students who went to study under him in Jena. The case of the hapless Russian student Jacques (Jakov) von Bedriaga suggests the extent to which Haeckel's students were picking up on polemics as the *modus operandi* of the German zoological community. In 1874, at the age of nineteen, Bedriaga had become embroiled in a priority dispute with Professor Theodor Eimer of Tübingen that dragged on for at least eight years. In justifying the polemical attacks he had made on Eimer, Bedriaga wrote in 1882, "In no other country are [polemical] replies so much the order of the day, as in Germany. By replying to the Tübingen professor, I have demonstrated that I have adapted myself to the customs of the country." He went on: "It is known everywhere to be an allowed thing in polemical writings to attack one's opponent sharply and bitingly, so long as the attack is clothed in a decorous form; also, no right-thinking person will feel personally offended by such modes of attack." In contrast to himself, Eimer had violated the bounds of proper polemicizing by accusing Bedriaga of base and vulgar motives, and by maligning the faculty of Jena for letting him get away with it.[50] If "biting" polemical attacks could be considered "decorous," and even "the order of the day," it is no wonder that Siebold was appalled by the behavior of the younger generation of zoologists!

Although the institute-based journals offered a ready outlet for disputatious remarks, the most important forum for establishing and attacking priority claims was a different sort of journal, the *Zoologischer Anzeiger*. Following a round of letters to solicit support in early 1878, the comparative anatomist Julius Victor Carus (brother-in-law of the publisher Wilhelm Engelmann) put out the first issue that summer. It was published twice a month to permit rapid printing of results and claims, it dated contributions, and it accepted short notices. It also brought out a running list of new titles of interest to zoologists, both monographs and articles. Finally, it published news of the profession. Although such "Fachblätter" existed in such other disciplines as botany and chemistry, zoologists had not heretofore had this kind of information gathered all in one place.

The *Anzeiger* was the perfect place for disputes over priority, giving them a legitimacy they had not enjoyed before. A pamphlet might or might not be bought or read by a colleague and might easily be dismissed as a public expression of things that ought to be left private. Even an institute-

---

50. J.V. Bedriaga, "Zweite Erwiderung an Herrn Th. Eimer," *Archiv für Naturgeschichte* 48 (1882): 303–308.

based journal ran the decided risk of losing credibility by laying bare the personal likes and dislikes of its editor. The *Zoologischer Anzeiger* was different. It took no sides, but allowed all involved parties their piece. Disputes in its pages were real conversations (or at least shouting matches), not rival monologues by individuals who weren't speaking to each other. Furthermore, because its other useful services gave it a very wide circulation, the *Anzeiger* also provided one of the largest audiences for such disputes. (Its central place in the discipline was secured in 1891, when it became the organ of the newly founded German Zoological Society.) By its evenhandedness and breadth of circulation, the *Anzeiger* went a long way toward acknowledging priority disputes as a legitimate part of scientific activity within the zoological community.

Finally, the last new journal deserving mention here was the *Biologisches Zentralblatt*, founded in 1881 under the coeditorship of three professors in Erlangen: the botanist Maximilian Reess, the physiologist Isidor Rosenthal, and the zoologist Emil Selenka. Like the *Anzeiger,* this journal was not primarily aimed at publishing the detailed empirical reports characteristic of a research journal. Instead, it attempted to provide its readers with an overview of the three neighboring areas of biology represented by its editors. To do this, they published original articles of general interest ("purely polemical articles will be . . . completely excluded"), summaries of important works, comprehensive overviews of problem areas, book reviews, and short notices.[51] With its emphasis on issues of interest to more than one specialty, the *Biologisches Zentralblatt* put theoretical and methodological discussions in the spotlight. Empirical results tended to be summarized very briefly, while the significance authors assigned to these results—and the arguments among various authors over how to weigh the significance—received far greater attention. This journal too appears to have had no trouble finding an audience or contributors, and despite the note about polemics, it certainly contributed its share to the general atmosphere of controversy.

So where did all this publishing activity leave the *Zeitschrift für wissenschaftlich Zoologie*? In the 1880s it continued to publish the same kinds of articles it had in the past: longish empirical discussions of research findings, accompanied by numerous illustrative plates. Where theoretical issues were addressed (and most authors did present their work as contributing something to the theoretical issues of the day), they usually came at the end,

51. "An unsere Leser," *Biologisches Centralblatt* 1 (1881–82): 1.

presented as implications drawn from a solid body of empirical research. No change here. Moreover, the journal drew from roughly the same constituency—it had always been primarily the domain of men in their twenties and early thirties, most frequently the students of the editors and their friends or long-term *Mitarbeiter*. As was the case earlier, the first work of a young scientist was usually submitted by his supervisor along with a letter attesting that the work had been carefully carried out in his institute under his own watchful eye (or, in the case of larger institutes such as Würzburg's, under the eye of the well-credentialed scientist who supervised the lab). And, following the established custom, many of these works were identical with the students' doctoral dissertations.

Nor was the most frequent sort of article appearing in the *Zeitschrift* in the 1880s much different from those of twenty years earlier. Typically, a person would select a bug, worm, or other small creature whose anatomy and development had received "inadequate attention" and then present a complete rundown of its various structures, describing its hard parts (shell or bone if there were any), circulatory and respiratory systems, musculature, nervous system, mode of reproduction, and cycle of development. Or, with a better-studied creature, one might concentrate on that system or structure which had been least studied. In any event, one was adding an empirical building block to the zoological "house of knowledge" (to borrow a frequent German phrase).

Despite the continuities, the role of the *Zeitschrift* in the community was not exactly what it had been earlier. In the 1850s and 1860s the publication of empirical research on the comparative anatomy, histology, reproduction, and development of invertebrates was part of a new program of research viewed by its participants as the forefront of zoological activity. By the 1870s and 1880s this program had become so well entrenched that the production of such publications had become part of the *training* of young zoologists. The very same sort of publication that had once served to advance the forefront of research now served primarily to demonstrate mastery of essential zoological skills: literature searching, use of the microscope, observing, and drawing.

By the 1880s, then, publishing an empirical research article in the *Zeitschrift* had gained a different meaning than it had had earlier. Rather than signifying that one was allying oneself with a particular research program, it meant that one was demonstrating one's basic competence as a professional researcher in zoology. There were still a few contributors who supplied the journal with articles over a relatively long period of time, but the vast majority of authors published only one or two articles, usually at the earliest stages of their careers.[52] The areas of ferment in the zoological community had moved elsewhere and found their expression in a different form. The presentation of one's discoveries in the form of the detailed research report was still necessary for certification, but if those discoveries engaged the community, the arguments over their originality and significance would take place in forums of open combat such as the *Zoologischer Anzeiger* or the *Biologisches Zentralblatt*, where the facts themselves often receded in the discussion of the ideas and personalities of the scientists.

In addition to expanding the number and kinds of publication opportunities for zoologists, the rash of new journals also helped erode the sense of community that had been so important to Siebold. The very growth in the number of journals available to zoologists helped break down that outlook of corporate unity, centered as it had been around one journal and one common idea of the proper aims and methods of zoology. But Siebold's conception of community was breached in another way by these journals as well, by their sanctioning of conflict and mistrust as accepted norms of scientific behavior. This marked a major shift in the character of the community.

Conflict and Community

"It is lamentable," wrote Weismann to Siebold in 1877, "that talent and purity of character are not always united, but it is so!"[53] In the scientific world Siebold sought to maintain, talent and character *did* go together. There was more than gentlemanly behavior at stake here; science itself was at issue. For Siebold, true and good science was founded on a base of *facts*. (Here it is telling that, like Darwin, who was only five years younger than he, Siebold spent decades painstakingly developing evidence for the great theory of his life—in Siebold's case, the theory of parthenogenesis, or reproduction by the female without fertilization from the male). The truths about nature, in Siebold's view, could only be found through the tedious, self-abnegating process of collecting results and subjecting them to one's

52. In volumes 31 to 45 (1878–1887), over 80 percent of the authors wrote only one or two articles, as compared with about 65 percent for volumes 16 to 30 (1866–1878). Among these authors, the large majority thank their "honored teachers" at the university institutes where they conducted their research, a fact that reinforces the impression that these were students' works.

53. Weismann to Siebold, 13 August 1877, EE 2092.

own skeptical and persistent scrutiny. This was not naive Baconianism—certainly theories might temporarily guide one's research, and one might even find a theory (such as parthenogenesis) that was upheld by the research. But in the search for nature's truths, the injection of the personal and subjective must be held to a minimum. Speculative theorizing, priority disputes, and polemical debates all served only to bury nature itself under layers of human vanity. None of them properly belonged to science. *Certainly* none of them belonged to the published record of scientific results.

Through the *Zeitschrift* Siebold sought to bring his scientific ideal as near as possible to reality. By accepting only empirically based research articles, he made the journal's published contents an implicit statement about the appropriate aims of scientific research. Moreover, his system of depending on recommendations and personal knowledge of contributors to his journal helped ensure (though it did not guarantee) that those scientists accepted into the fold would already know the norms of scientific behavior and follow them.

As we have seen, however, the very sorts of disputatious behavior that Siebold most objected to, and most objected to sanctioning by making public, came to form a central part of the scientific activity of the late 1870s and early 1880s. Once polemics and priority disputes began to appear with regularity in journals, they clearly had the sanction of the community. Why did this come to pass?

Traditional histories of biology discuss the tide of evolutionary speculation that swept over Germany in the late nineteenth century, and one might be inclined to associate the polemics of the period with the conflict between empiricists such as Siebold and speculative theorists such as Haeckel. There is some justice to this point of view; the extent to which theories needed to be founded in scientific evidence was clearly a subject for serious methodological and epistemological debate. But this analysis does not seem to account for the intensity of the debates or the place of priority disputes within them.

A more satisfactory explanation, drawn from the sociology of science, might look to scientific careerism. The aggressive pursuit of their own career advancement is one of the principle characteristics of the zoologists active in the period, and it not only provides a motive for the priority disputes and the intensity of the polemical debates but also helps explain the shift in role of the *Zeitschrift für wissenschaftliche Zoologie* within the community.

One can see this attention to career interests in the contrast a younger zoologist drew between Siebold's attitude toward research and that of his own generation. In an eloge for Siebold written in 1886, Richard Hertwig, his successor to the chair of zoology at Munich, described Siebold as writing about whatever struck his fancy. Even when he was researching parthenogenesis, Siebold did not focus narrowly on that aspect of the bees and wasps he was studying but also took note of other things, such as how they made their nests and how they protected themselves against enemies. Somewhat nostalgically, Hertwig went on to point out that zoologists of his day didn't have the luxury of this "undirected [*absichstslos*] manner of research," but had rather to focus closely on a particular problem and get it solved (and, one might add, get the solution published). Hertwig himself had displayed such career-mindedness early on, writing to Claus in 1878 that since he was almost finished with his current project, he was "proceeding from the principle that younger zoologists must use their time to gain an overview over the animal kingdom from their own investigations"; accordingly, he was looking for a new organism to work on.[54]

The intense concern to establish one's professional credentials seems not to have overtaken zoology until the late 1870s and early 1880s. Given that natural history is a very old subject, this might be surprising, but in fact zoology gained independent status as a university discipline in Germany only gradually in the nineteenth century, and it was not until the late 1860s that very many people could think of making a full-time career as a zoologist.[55] The generation establishing their careers then was that born in the mid-1830s, which included Claus, Weismann, Semper, and Haeckel, four of the leading disputants of the 1870s and 1880s, as well as Ehlers, who stood at the center of the storm without contributing much directly to it. All of them had achieved professorships before the disputes became very heated, it is true, but contemporary letters and university records about hiring decisions in the late 1870s indicate that they were still greatly involved in competing for more prestigious chairs and larger institutes.[56] As Hertwig's comments suggest, these scientists certainly instilled such thinking in their students, the generation born around 1850. For this generation, which

54. Richard Hertwig to Carl Claus, 19 August 1878, (Bl. 11–12), BSPK Lc 1898 (29), R. Hertwig. Hertwig's contemporary Hubert Ludwig expressed a similar sentiment in a letter to Ehlers on 12 December 1878. EE 1158 (No. 11).

55. Lynn Nyhart, "The Disciplinary Breakdown of German Morphology, 1870–1900," *Isis* 78 (1987): 365–389, especially 370–371.

56. See Lynn K. Nyhart, "Morphology and the German University, 1860–1900" (Ph.D. diss., University of Pennsylvania, 1986), especially chaps. 2 and 3. I plan to expand the discussion of zoology's disciplinary development in my book (currently in progress).

included such men as Hertwig, Otto Bütschli, and Wilhelm Roux, the competition for jobs was even tighter than the competition for prestige and autonomy had been for their teachers.[57]

Career-minded thinking is similarly evident in the sorts of publishing these men were doing. Priority disputes, the source of the ungentlemanly behavior evinced by Claus, Weismann, and so many others, are disputes about credit or recognition, which is the central token of achievement in a system built around originality. The *Zoologischer Anzeiger*'s timely publication of hurried "preliminary results" and its public communication of priority disputes were every bit as much responses to the needs of aspiring professionals as its running list of recent publications and its advertisements for assistantships. Similarly, as we have seen, the *Zeitschrift*'s empirical research articles also represented a building block, not just in scientific knowledge but in a scientific career, by demonstrating the young zoologist's technical competence.

All of these career-related activities fit nicely into well accepted sociological explanations about how "the scientific community" operates. Yet somehow this analysis still does not seem to account for why in this *particular* community polemics and priority disputes became so much, as Jacques Bedriaga put it, "the order of the day."

I suggest that these priority disputes were not just about priority.[58] As important as credit for one's contributions is among scientists—and I am persuaded that this credit is the currency of exchange within any scientific community—the need for it does not seem to me to warrant simultaneous engagement of the members of an entire community in bitter disputes over getting their due. And it is scarcely an exaggeration to say that to be a zoologist in Germany around 1880 was to be involved in such disputes.

Could it be that as the *Gemeinschaft* Siebold had in mind turned into the *Gesellschaft* of "professional zoology," the priority dispute itself served to assert one's membership in the community? By attacking someone else, a marginal scientist might be asserting his right to be heard, in part by

57. The details of this competition for jobs and the underlying causes of the rapid increase in the number of aspiring zoologists in the 1870s cannot be explored here, but I hope to publish a separate work on this in the future.

58. As Augustine Brannigan and Michael Mulkay have stressed, disagreements labeled "priority disputes" frequently concern not just who was the first to make an observation, but also its very meaning. See Augustine Brannigan, *The Social Basis of Scientific Discoveries* (Cambridge: Cambridge University Press, 1981); Michael Mulkay, *The Word and the World: Explorations in the Form of Sociological Analysis* (London: George Allen and Unwin, 1985), 191. Although this is also the case for many of the disputes in this community, my point here is somewhat different.

presenting himself as having competence at least equal to the other person. (This stance is illustrated by Bedriaga, who, it will be recalled, took on a professor at the age of nineteen.) But these disputes, as we have seen, did not take place only among marginal scientists, and this explanation accounts at best for only a portion of the disputes. I would like to suggest a further way in which these disputes served their participants' social interests.

As the form of the priority dispute took shape in this community, the claims made took on certain generic elements. It did not seem to matter, after the first claim was made, which man had accused the other of stealing. If the dispute extended beyond one exchange, these generic elements entered into the arguments of both sides. As the controversy between Claus and Weismann illustrates, the participating zoologist would assert not only his claims to credit, but also his integrity, his originality, his allegiance to "the facts of the matter," even his reluctance to engage in disputes, while at the same time demonstrating his persuasive skills in casting doubt on the possession of those same qualities by his opponent.

These are the very qualities that Siebold held dear and that are commonly held within a scientific community. The difference is that in the community Siebold sought to perpetuate through the *Zeitschrift*, none of this public breastbeating was necessary—the acceptance of these values was a part of the quiet process by which a scientist was let into the ranks of *Mitarbeiter*. But as the older community ideal disintegrated, it became increasingly necessary for the members of the community to assert their membership and their allegiance to its values in public. Publishing empirical research reports in journals like the *Zeitschrift* was still one way of doing this; ironically, engaging in a priority dispute was another.

Research for this article was carried out under summer grants from the American Council of Learned Societies and the National Endowment for the Humanities. The author gratefully acknowledges their support.

Bruce J. Hunt

3. Rigorous Discipline: Oliver Heaviside Versus the Mathematicians

The following story is true. There was a little boy, and his father
said, "Do try to be like other people. Don't frown." And he tried
and tried, but could not. So his father beat him with a strap; and
then he was eaten up by lions.

Reader, if young, take warning by his sad life and death. For
though it may be an honour to be different from other people, if
Carlyle's dictum about the 30 millions be still true, yet other people
do not like it. So, if you are different, you had better hide it, and
pretend to be solemn and wooden-headed. Until you make your
fortune. For most wooden-headed people worship money; and,
really, I do not see what else they can do. In particular, if you are
going to write a book, remember the wooden-headed. So be rigorous; that will cover a multitude of sins. And do not frown.

Oliver Heaviside, *Electromagnetic Theory,* Volume 3, page 1.

IN MAY 1894, Oliver Heaviside sent the Royal Society the third installment
of his long paper "On Operators in Physical Mathematics." Although he
was largely self-educated and held no official position, Heaviside was recognized as one of the leading electromagnetic theorists in Britain and had
been elected to the Royal Society in 1891.[1] Like all Fellows of the Society, he
was allowed to publish papers in its *Proceedings* without going through
formal refereeing, and the first two installments of his paper on operational

1. For biographical information, see Paul J. Nahin, *Oliver Heaviside: Sage in Solitude* (New
York: IEEE Press, 1988). In subsequent notes, the following abbreviations are used: RS: Royal
Society of London; OH-IEE: Oliver Heaviside Papers, Institution of Electrical Engineers,
London; FG-RDS: G. F. FitzGerald Papers, Royal Dublin Society; *Proc. RS: Proceedings of the
Royal Society of London; EMT:* Oliver Heaviside, *Electromagnetic Theory.* 3 vols. (London: The
Electrician Co., 1893–1912; reprint New York: Chelsea, 1971; original dates of publication of
articles that first appeared in the *Electrician* are given in brackets).

methods and infinite series had appeared without incident in 1893.[2] But the third part was not so fortunate: two months after sending it in, Heaviside received a brief form letter from Lord Rayleigh, the Secretary of the Royal Society, stating, that "the Committee of Papers, not thinking it expedient to publish it at present, has directed your manuscript to be deposited in the Archives of the Society." Rayleigh added in a private note that the paper "does not commend itself to our mathematicians"; they regarded it as "an attempt to do by imperfect methods what has already been done by rigorous ones," and as a breach of mathematical standards serious enough to merit suspension of the custom of free publication by Fellows.[3] Heaviside was understandably upset at having his work rejected in this way, and he lost few opportunities in later years to vent his bitterness at the "rigorists" of the Royal Society. He managed to retrieve his manuscript from the Archives and included parts of it in the second volume of his *Electromagnetic Theory* in 1899, but the full text of Part III has never been published. The manuscript was later badly damaged, and its remains now lie tattered and partly illegible among the collection of Heaviside's papers held by the Institution of Electrical Engineers in London.[4]

The story might well have ended there; Heaviside was hardly the first person to have had a paper rejected, even if the circumstances in his case were a little unusual. But in the 1910s and 1920s his operational work began to attract new attention, both from electrical engineers interested in its application to circuit problems and from mathematicians intent on discovering why it worked as well as it did; physicists, too, began to draw on it in connection with Paul Dirac's new formulation of quantum mechanics.[5] As Heaviside's methods came into widespread use, and as ways were found to put most of his results on a basis that even the strictest rigorist could accept, the rejection of his 1894 paper was increasingly called into question.

This later rehabilitation of Heaviside's methods, together with the biting style of his attacks on the rigorists, has given rise to a folklore—really several folklores—in which he is depicted to electrical engineers and physi-

2. Oliver Heaviside, "On Operators in Physical Mathematics, Part I," *Proc. RS* 52 (2 February 1893): 504–529, and "Part II," *Proc. RS* 54 (15 June 1893): 105–143.
3. Rayleigh to Heaviside, 26 July 1894, OH-IEE.
4. Heaviside, *EMT* 2 (7 October, 2 December 1898): 457–482 is described on page 457 as a version of Part III condensed to less than a third of its original length. The badly damaged manuscript of Part III was reassembled by Eleanor Symons and is now in Box 14, OH-IEE.
5. Nahin (n. 1), 227–230, 304; on the use of Heaviside's operators in quantum physics, see Max Jammer, *The Conceptual Development of Quantum Mechanics* (New York: McGraw-Hill, 1966), 227–228. As Jammer notes, Dirac learned Heaviside's methods during his undergraduate training in electrical engineering.

cists as a persecuted genius and to mathematicians as a brilliant but mis-guided amateur. Historians of mathematics have cleared up some priority questions involving Heaviside's predecessors and have shed considerable light on the later attempts to "rigorize" his operational calculus, but they have so far paid relatively little attention to the circumstances of the 1894 rejection itself or to what this episode and its aftermath can tell us about disciplinary practices in mathematics.[6] In what follows, I will use Heavi-side's conflict with the mathematicians as a window into a little-explored issue: the *rhetorical* dimension of mathematical argumentation, in particu-lar the way claims about "rigor" have been used to maintain and enforce disciplinary boundaries.

Pairing "rhetoric" with "mathematics" may at first seem an odd or even contradictory move. As Philip J. Davis and Reuben Hersh noted in a recent article on the subject,

> If rhetoric is the art of persuasion, then mathematics may seem to be its antithesis. This is believed, not because mathematics does not persuade, but rather because it seemingly needs no art to perform its persuasion. The matter does it all; the manner need only let the matter speak for itself.[7]

But such a strict separation of mathematics from "mere rhetoric" does not long survive an examination of the history of mathematics, in particular of the shifts in standards of rigor that have repeatedly redirected mathematical work. Proofs that seemed perfectly conclusive to Euler in 1750 were rejected as unrigorous by Cauchy in 1825, and Cauchy's proofs were in turn judged inadequate by Weierstrass in 1870.[8] Conversely, the rigor demanded by the Greek geometers was quite foreign to early modern mathematicians. In shifting what they have counted as constituting a valid argument, mathe-maticians have, in effect, shifted the rhetorical practices that they have

6. Jesper Lützen, "Heaviside's Operational Calculus and the Attempts to Rigorise It," *Archive for History of Exact Sciences* 21 (1979): 161–200; S. S. Petrova, "Heaviside and the Development of the Symbolic Calculus," *Archive for History of Exact Sciences* 37 (1987): 1–23. The only extended discussion of the 1894 rejection itself is in J. L. B. Cooper, "Heaviside and the Operational Calculus," *Mathematical Gazette* 36 (1952): 5–19, and aims mainly at defending the rejection of the paper; Nahin (n. 1), 222–226, follows Cooper's account fairly closely.

7. Philip J. Davis and Reuben Hersh, "Rhetoric and Mathematics," in John S. Nelson, Allan Megill, and Donald N. McCloskey, eds., *The Rhetoric of the Human Sciences* (Madison: University of Wisconsin Press, 1987), 53–68, on 53. The quoted remark is not meant to reflect the views of Davis and Hersh themselves, but is their summary of the conventional wisdom.

8. See Philip Kitcher, *The Nature of Mathematical Knowledge* (New York: Oxford University Press, 1984), 241–271, and Morris Kline, *Mathematical Thought from Ancient to Modern Times* (New York: Oxford University Press, 1972), 426–434, 947–978.

agreed to accept as appropriate and convincing. By paying close attention to the rhetorical strategies that were successful in particular circumstances—and to those that failed—we can gain important insights into the aims and interests of the mathematicians of a given time and place. In particular, by examining how mathematicians used the rhetoric of "rigor" to draw and redraw the boundary separating their own discipline from physics and engineering, we can better understand what was really at stake, and why mathematicians responded as they did, when Heaviside unknowingly crossed that line in 1894.

Telegraph Operators

Heaviside invented his operational calculus in the 1880s as a tool for the analysis of telegraphic circuits, and although he and others later turned it into a sophisticated and seemingly abstract system, it remained rooted in the practical needs of electrical technology. Heaviside first came into contact with that technology in 1868, when at the age of eighteen he started work as a telegrapher on the new Anglo-Danish cable, first in Denmark and later at Newcastle. He had grown up poor in London and left school when he was sixteen. A fortunate family connection with Charles Wheatstone, the inventor of the telegraph, led to his job on the cable, and from then on telegraphy formed the core of all of Heaviside's work.

Heaviside was not content merely to operate the telegraph; he wanted to understand it as well, and soon set about analyzing telegraphic phenomena mathematically. He had great mathematical talent and managed to teach himself an enormous amount of mathematics and physics while at Newcastle, mostly from such standard works as Isaac Todhunter's *Differential Calculus* and William Thomson and P. G. Tait's famous *Treatise on Natural Philosophy*. From these he absorbed the techniques and standards of mid-nineteenth-century British (that is, mainly Cambridge) mathematics, which was then far more concerned with finding solutions to physical problems than with questions of "pure mathematics" or abstract rigor. Heaviside found this approach congenial, and even at the height of his later attacks on the "rigorists" of the 1890s, he made a point of defending the older school of Cambridge mathematicians.[9]

Heaviside left Newcastle in 1874, in part because of health problems, and

9. Heaviside, *EMT* 2 (14 December 1894): 10–11.

returned to live with his parents in London. Although he never again held a regular job, he continued to study telegraphic problems on his own, often in considerable poverty. He turned out huge amounts of work and soon began to publish regularly in the *Philosophical Magazine*, the *Journal of the Society of Telegraph Engineers*, and, after 1882, in the *Electrician*, a weekly London trade journal. He focused mainly on the mathematical analysis of telegraphic propagation and in the mid-1880s became one of the first to apply Maxwell's theory of electromagnetic waves to such phenomena. By the end of the decade he had emerged as one of the leading interpreters of Maxwell's theory, and it was in fact Heaviside who first cast it into the compact form now universally known as "Maxwell's equations."[10]

Heaviside called his work "physical mathematics"; he told Heinrich Hertz in 1889 that one should always "keep as near to the physics of the matter as one can, and not be deluded by mere mathematical functions," and later declared that "Physics is above mathematics, and the slave must be trained to work to suit the master's convenience."[11] His work on vector analysis, on Maxwell's equations, and on the operational calculus itself all reflected Heaviside's conviction that mathematics was primarily a language for the expression of physical relationships, and that it was the task of the "physical mathematician" to devise the most appropriate language for the case at hand. His attitudes were those of a physicist or an engineer; "pure mathematics," divorced from physical applications, held little interest for him, and he had a positive distaste for abstract exercises of the "logic-chopping" kind.[12]

Heaviside's first really advanced mathematical work dates from the mid-1870s, when he tackled the theory William Thomson (later Lord Kelvin) had devised in 1854 to describe the propagation of signals along a submarine cable. The differential equations governing the variation of voltage and current along a cable are complicated, but Thomson was able to solve the simpler cases by using the technique of expansion in trigonometric series that Joseph Fourier had developed nearly fifty years before to treat the propagation of heat.[13] As Heaviside extended and generalized Thom-

10. See my dissertation, "The Maxwellians" (Johns Hopkins University, 1984), and my forthcoming book (Cornell University Press).

11. Heaviside to Hertz, 13 July 1889, quoted in J. G. O'Hara and W. Pricha, *Hertz and the Maxwellians* (London: Peter Peregrinus, 1987), 67; Heaviside, *EMT* 2 (14 January 1898):414.

12. See Heaviside, "The Teaching of Mathematics," *Nature* 62 (4 October 1900): 548–549; reprint Heaviside, *EMT* 3: 513–514. Heaviside's distaste for the deductive approach may explain his poor performance in Euclidean geometry at school; see Nahin (n. 1), 17.

13. William Thomson, "On the Theory of the Electric Telegraph," *Proc. RS* 7 (May 1855): 382–399; reprinted in Thomson, *Mathematical and Physical Papers*, vol. 2 (Cambridge: Cambridge University Press, 1884), 61–76. The paper consists mainly of two letters Thomson sent G. G. Stokes in October 1854 and Stokes's reply.

son's theory in the 1870s and 1880s, he became a master of Fourier analysis; indeed, he later told his friend G. F. FitzGerald that he "practically re-discovered all Fourier's mathematics and a lot more too" in the course of his telegraphic work, starting only from "hints or references" in the few books available to him.[14] Heaviside's lack of books and of formal education often left him ignorant of what others had already accomplished, but by forcing him to work out so much for himself, it gave him an exploratory approach to mathematics that sometimes led him to results of striking originality and value.

Not all circuit problems yielded readily to Fourier analysis and other "classic" techniques, and it was to attack these cases that Heaviside devised his operational calculus in the 1880s.[15] He began by observing that in a simple resistive circuit we can regard the resistance R as an "operator" that turns the current C into the voltage V, in this case by simple multiplication ($V = RC$). Similarly, in a more complicated circuit, we can construct a "resistance operator" Z that will turn a given current into the correspond-ing voltage ($V = ZC$), generally by some combination of differentiations and integrations. The resistance operator of a circuit can be found by combining the resistances, capacitances, and inductances according to a few simple rules, but solving the resulting differential equation to find the voltage and current as a function of time is far more difficult. As the energy of the circuit sloshes back and forth between the condensers and the coils, the voltage and current oscillate and subside in a complicated tangle of characteristic "normal modes." The beauty of Heaviside's operational method was that it enabled him to bypass the general differential equation of the circuit—which was often almost impossible to solve, particularly with all of the proper boundary conditions—and calculate the actual be-havior of the circuit directly from the form of its resistance operator.

Heaviside built his operational calculus from a number of pieces, not all of which he invented himself. One of its best-known features, the "symboli-cal" treatment of operators, was in fact quite old; mathematicians had been manipulating symbols for differentiation and integration as if they were ordinary algebraic quantities since Leibniz's time, and Heaviside learned the technique from George Boole's 1859 *Treatise on Differential Equations*.[16]

14. Heaviside to FitzGerald, 26 February 1894, FG-RDS.
15. Heaviside began using operational techniques around 1881, but he first discussed them at length in a paper, "On Resistance and Conductance Operators and their Derivatives, Inductance and Permittance, especially in connection with Electric and Magnetic Energy," *Philosophical Magazine* 24 (December 1887): 479–499; reprint Heaviside, *Electrical Papers* (London: Macmillan, 1892; New York: Chelsea, 1970), vol. 2, 355–374.
16. Jammer (n. 5), 224–227; Petrova (n. 6).

Heaviside made particular use of the time differentiation operator, d/dt, which he abbreviated as "p," and manipulated it and other detached operators very freely, rarely pausing to question their meaning or application as long as they could be made to yield useful answers.

Perhaps the most powerful of Heaviside's techniques was one he took directly from practical telegraphy. When a telegraph key is pressed down, the voltage at the terminal suddenly jumps from zero to some steady value, and the current then responds in a way characteristic of the circuit. This jump can be represented mathematically by a "step function" (0 before time t = 0 and 1 thereafter), which Heaviside wrote simply as "**1**." Thomson had used the idea in his 1854 cable theory, but Heaviside was the first to exploit its full power. Applying a sudden jolt like that represented by a step function excites the inherent normal modes of oscillation of a circuit, just as ringing a bell excites its various tones. If the relative amplitudes and decay rates of the different normal modes can then be found, the resultant behavior of the circuit follows by simple addition. Heaviside solved this very difficult problem in 1886 by using his knowledge of the energy relations of electric circuits to derive an "expansion theorem" that gave the amplitudes and decay rates directly from $Z(p)$, the resistance operator as a function of the time differentiator.[17] Along with related techniques involving Laplace transforms, the expansion theorem has been widely used by electrical engineers since the 1920s to analyze the transient response of circuits.[18] Heaviside's derivation of the theorem from electromagnetic considerations was typical of his approach to "physical mathematics," but he later showed that it applied to a much wider class of linear differential equations as well.

By this time Heaviside had already taken several steps that a strict mathematician might question, but he soon ventured even deeper into suspect territory. In treating distributed circuits, particularly cables, Heaviside's formalism often led him to fractional powers of p. But what meaning could be assigned to "$p^{1/2}$"? How could one take the square root of differentiation, or of any operation? Even Heaviside was puzzled, and admitted that "according to ordinary notions," such fractional differentiation was "unintelligible."[19] But unintelligibility was no obstacle to Heavi-

17. Heaviside's rather complicated path to this result is recounted in Lützen (n. 6), 193–198.

18. Perhaps the clearest account of how Heaviside's methods were used by engineers is Vannevar Bush, *Operational Circuit Analysis* (New York: Wiley, 1929); on the relationship between Heaviside's expansion theorem and the "Laplace transform" methods that took its place after the 1930s, see Balthasar van der Pol and H. Bremmer, *Operational Calculus Based on the Two-sided Laplace Integral* (Cambridge: Cambridge University Press, 1955).

19. Heaviside, *EMT* 2 (26 June 1896): 286.

side, and by comparing his operational solutions with those found by other methods, he soon found a workable way to handle fractional differentiation and to use it in solving cable problems. He even derived the curious but very useful result that $p^{1/2}1 = (\pi t)^{-1/2}$; that is, the square root of differentiation applied to a unit step function equals $1/\sqrt{\pi t}$. This was in fact not a new result, although Heaviside had discovered it on his own. He knew from a brief reference in Thomson and Tait's *Treatise* that mathematicians had investigated fractional differentiation in the eighteenth and early nineteenth centuries, but having no access to the relevant publications, he was forced to work it all out for himself when he needed it. Much like the mathematicians of two or three generations before, he approached the subject experimentally, extending his formalism as far as it would go and checking his results against those found in other ways, with no pretense of giving strict proofs. He did not know that work on fractional differentiation had run into contradictions in the 1840s and 1850s, or that the whole subject had subsequently fallen into disrepute among mathematicians, but it seems unlikely that such knowledge would have deterred him from using a tool that he had found to be of practical value.[20]

Heaviside followed a similar pattern in his work on divergent series. His operational solutions often took the form of infinite series, and although many of these converged to finite values, others did not, either oscillating wildly from large negative to large positive values or simply growing term by term without limit. Such divergent series had been widely used in the eighteenth and early nineteenth centuries, but much like fractional differentiation, they had given rise to so many contradictions that mathematicians eventually ruled them out of bounds; indeed, Niels Abel had declared in 1826 that divergent series were "the invention of the devil" and should be banished from serious mathematics.[21] Some mention of them survived in the literature, however, and Heaviside learned from works by Stokes and Boole that a very good approximate value of a function could often be obtained by taking the first few terms of its divergent expansion. This was "eye-opening," he later said, and led him to modify his earlier and quite orthodox belief that divergent series were simply meaningless.[22] After refining and extending the approximation technique, Heaviside soon found ways to use divergent series of operators to generate convergent algebraic

20. Heaviside (n. 2), "Part I," 516; see also Bertram Ross, "The Development of Fractional Calculus, 1695–1900," *Historia Mathematica* 4 (1977): 75–89.

21. Quoted in Kline (n. 8), p. 973.

22. Heaviside (n. 2), "Part II," 121–122.

series and to solve other numerical and analytical problems. He became convinced that divergent series were of fundamental importance, and despite the obvious difficulties in handling expressions that gave every appearance of being nonsense, he began to use them quite freely.

How did Heaviside justify his many mathematical leaps? Very simply: they worked. By treating operators as if they were ordinary numbers, by extending differentiation to fractional values, and by manipulating divergent series as if they had determinate values, he was able to solve, often in just a few lines, problems that were intractable by other methods. When he substituted his answers back into the original differential equations or checked them in other ways, they almost always turned out to be correct. Heaviside was admittedly unsure *why* his methods worked as well as they did, and he warned that they sometimes broke down. "No matter how sweetly the algebraical treatment of operators may work sometimes," he said, "it is subject at other times (owing to our ignorance) to the most flagrant failures"; serious errors could be avoided only by having a sure grasp of the physics of the problem or by checking the answer in other ways.[23] But the fact that his operational calculus usually *did* work proved, he thought, that it must embody some underlying mathematical truth, while its occasional failures simply meant that more remained to be discovered about it. Indeed, Heaviside declared that the interest of the subject, already very great, was only "heightened by the mystery that envelops certain parts of it."[24]

Up to the early 1890s, Heaviside's published work fell entirely within physics and electrical engineering. He framed his arguments and assembled his evidence in accordance with the standards of those disciplines, and the recognition he eventually began to receive, culminating in his election to the Royal Society in 1891, came mainly from physicists. In his work on fractional differentiation and divergent series, however, Heaviside often found himself reasoning not about voltages and currents but about number and quantity in the abstract. He discovered mathematical patterns and techniques that seemed to him both new and important, and believing that they deserved "to be thoroughly examined and elaborated by mathematicians," he decided to take advantage of the outlet provided by the Royal Society to bring his discoveries to their attention.[25] In doing so, however, he stepped across an invisible but very real disciplinary boundary, and he

23. Heaviside (n. 2), "Part I," 515.
24. Ibid., 506.
25. Ibid.

soon found that the mathematicians of the 1890s played by a set of rules that he neither understood nor accepted.

Drawing the Line

Heaviside sent Part I of his paper "On Operators in Physical Mathematics" to the Royal Society in December 1892, and it duly appeared in the *Proceedings* a few months later. The paper began with a brief sketch of the operational calculus as a whole, but its main focus was "the more transcendental matter" of fractional differentiation, including the manipulation of expressions involving $p^{1/2}$ and the construction of generalized forms of the factorial and exponential functions.[26] Heaviside admitted that he had been unable to consult any of the earlier works on the subject, but he did not consider this altogether regrettable; even if not entirely new, he said, his investigation "has at least the recommendation of having been worked out in a mind uncontaminated by the prejudices engendered by prior knowledge acquired at second hand."[27] It was not a remark calculated to appeal to better-read mathematicians, but Heaviside seemed confident that his paper contained enough valuable results to be able to stand on its own merits.

In Part II of his paper, which reached the Royal Society in June 1893 and was published very shortly thereafter, Heaviside applied his operational methods to zero-order Bessel functions and began an extended discussion of divergent series. Such series could, he said, be "equivalent" in three different ways—numerically, algebraically, and analytically—but he refused on principle to define these terms explicitly. "Believing in example rather than precept," he said,

> I have preferred to let the formulae, and the method of obtaining them, speak for themselves. Besides that, I could not give a satisfactory definition which I could feel sure would not require subsequent revision. Mathematics is an experimental science, and definitions do not come first, but later on. They make themselves, when the nature of the subject has developed itself. It would be absurd to lay down the law beforehand.[28]

This was an excellent summary of Heaviside's philosophy of mathematics, but again, it was not a remark likely to satisfy more orthodox mathemati-

26. Ibid., 512.
27. Ibid., 517.
28. Heaviside (n. 2), "Part II," 121.

cians. Instead of starting with explicit definitions and laying out of his conclusions in a deductive chain, Heaviside recounted the stages in the evolution of his own acceptance of divergent series and called on mathematicians to keep an open mind about them. He reminded them that imaginary numbers had been rejected as nonsensical for many years before they finally won acceptance (indeed, they were regarded with suspicion at Cambridge well into the nineteenth century) and suggested that divergent series might eventually follow the same path. "In any case," he said, "we should not be misled by apparent unintelligibility to ignore the subject," for it was only by striking out into unknown territory that real progress could be made.[29]

The third part of Heaviside's paper, on higher-order Bessel functions, gave him more trouble than the first two. He found a serious error in Part II a few months after it was published, and the need to correct it, and to make laborious numerical checks of his series expansions, was discouraging. "I have a nightmare of neglected duty, the task of writing Parts 3 and 4 of 'Operators in Physical Mathematics,'" he told FitzGerald in February 1894, "and I can't get started."[30] It took him several more months, but he finally completed Part III in May 1894 and sent it off to the Royal Society with every expectation that it would appear in the *Proceedings* in due course.

Such was not to be the case. One of our few sources on the events that followed is a brief but revealing account by E. T. Whittaker, then a young student at Cambridge and later a great admirer of Heaviside's work. Writing in 1928, Whittaker noted that Heaviside's free use of fractional operators and divergent series could not help but irritate "professional pure mathematicians," particularly those of the 1890s:

At that time the most influential of them were trying to raise the standard of "rigour"—to move away from the happy old easy-going Todhunter period to the style which the Germans had adopted under the influence of Weierstrass: and the "operation" of extracting the square root of the process of partial differentiation seemed worse than anything in Todhunter—a kind of mathematical blasphemy. Not long afterwards one of them told me what happened. "There was a sort of tradition," he said, "that a Fellow of the Royal Society could print almost anything he liked in the 'Proceedings' without being troubled by referees: but when Heaviside had published two papers on his symbolic methods, we felt that the line had to be drawn somewhere, so we put a stop to it."[31]

29. Ibid., 124.
30. Heaviside to FitzGerald, 26 February 1894, FG-RDS.
31. E. T. Whittaker, "Oliver Heaviside," *Bulletin of the Calcutta Mathematical Society* 20 (1928): 199–220; reprinted in Chelsea edition of Heaviside, *EMT* 1: xiii–xxxiv, on xxix–xxx.

It was apparently A. R. Forsyth, then emerging as one of the leaders of the new generation of Cambridge pure mathematicians, who "drew the line" when Heaviside's paper came before the Council of the Royal Society at its July 4 meeting.[32] Rayleigh and Oliver Lodge tried to fend off Forsyth's attack, but without much success, and instead of being "read" (in title only) and then going directly to the printer, as was usual with papers by Fellows, Part III of Heaviside's paper was referred to William Burnside, professor of mathematics at the Royal Naval College in Greenwich. With that, its fate was sealed.

Burnside had been trained at Cambridge toward the end of the "happy old easy-going Todhunter period," but after graduating as second wrangler in 1875 he had moved away from the "mixed" or applied mathematics long traditional at Cambridge and toward "pure mathematics" of the kind more common on the Continent. From the early 1890s he worked mainly on the theory of groups. He was elected to the Royal Society in 1893 and was often asked to referee papers both for it and for the London Mathematical Society. His own work was noted for its elegance and rigor, and he held others to the same standard. "He could not be called lenient," his friend Forsyth wrote in his obituary in 1928; indeed, "he could even be severe on occasion," and in later years blocked papers by A. N. Whitehead and Srinivasa Ramanujan, among many others.[33] If one were looking for a referee likely to kill a paper like Heaviside's, Burnside would be a good choice.

The referee report on Part III of "On Operators in Physical Mathematics" now in the files of the Royal Society is unsigned, but it is in William Burnside's handwriting and is marked "Heaviside by Burnside" on the back.[34] It bears no date, but was evidently written in July 1894. Although he had been asked to report only on Part III, Burnside made it clear that his remarks were meant to apply to Parts I and II as well. All three were riddled with errors, he said, and showed both ignorance of important earlier work

32. See Rayleigh to Oliver Lodge, 5 July 1894, Lodge Papers, University College London, saying, "I was sorry about Heaviside yesterday, but know no help for it," and Lodge to Rayleigh, 9 July 1894, Rayleigh Papers (microfilm), American Institute of Physics, saying that "a healthy robustious person like Forsyth may airily cause a tragedy without meaning it." The lesson, Lodge said, was that "It's a mistake to stray into the close presence of the pure mathematician." Forsyth and Lodge had been elected to the Council of the Royal Society in 1893.

33. A. R. Forsyth, "William Burnside," *Proc. RS* 117A (1928): xi–xxv, on xiv. See Burnside's scorching rejection of an 1899 paper by Whitehead, RR.14.269, RS Archives, and an undated letter from G. H. Hardy to Joseph Larmor, no. 687, Larmor Papers, RS Archives, complaining that Burnside had done "his level best to squash Ramanujan's paper."

34. "Heaviside by Burnside," RR.12.135, RS Archives. The following quotations from Burnside are all from this report.

and disregard for proper standards of rigor. Heaviside appeared to be "quite ignorant of the modern developments of the theory of linear differential equations," although the papers Lazarus Fuchs and others had published since the 1860s were directly relevant to the problems he sought to solve. As it was, Heaviside's paper amounted to little more than an attempt "to find a 'royal road' to results which have already been established by exact reasoning," Burnside said, and so had no real value.

Burnside directed his strongest fire not, as Whittaker and others later implied, at Heaviside's operational method itself, but at his use of divergent series. The elimination of such series from serious mathematics had been one of the main triumphs of the movement for greater rigor begun earlier in the century by Cauchy and Abel. Burnside regarded it as a settled fact that divergent series were "analytically meaningless," and he found Heaviside's use of them outrageous. "Almost every paragraph of the paper will give an instance of an essentially divergent series being used as though it were a convergent one," he said, and in the absence of any apparent justification for "this entirely new departure," he could only conclude that Heaviside's results were nonsense. "Detailed criticism of results obtained in this way seems out of place," Burnside said. "They may or may not be true, but the way in which they are arrived at makes them absolutely valueless."

This was the central issue: Did the admittedly loose and exploratory way Heaviside had arrived at his results indeed render them "absolutely valueless"? Or were there other ways besides strict logical proof by which the reliability and value of mathematical results could be established? Like most mathematicians, Burnside regarded mathematics as purely deductive, and held that the truth of a mathematical proposition could be ensured only by attending to the rigor of its derivation. Even as he conceded that some of Heaviside's numerical approximations were "interesting in themselves and may possibly be of practical value," he denied absolutely that they could be used, as Heaviside claimed, to establish the "equivalence" of divergent series. Since Heaviside had not derived his results about such things as fractional differentiation and divergent series by logical steps from accepted premises, his work was little better than gibberish. Put another way, since he had not followed the established rules for the construction of a mathematical argument, he had failed to produce conclusions that a mathematician could find convincing.

Heaviside, of course, took a very different view. He regarded mathematics as essentially empirical, differing only in degree from physics and the other natural sciences; we accept its propositions, he said, not primarily because they have been derived by strict deduction, but because they work in practice and fit with our experience. Indeed, he asked, how can we obtain a starting point for our deductions except from experience?[35] He did not object to rigorous proofs when they could be had, but he did not regard them as the only valid route to mathematical knowledge. The principles of the operational calculus had been discovered by a combination of bold generalization and careful empirical testing; the fact that they had not been derived by strict logic did not mean that they were not true, or even that there were not excellent grounds for accepting them. Heaviside regarded mathematics primarily as a tool for solving problems, and as long as his techniques gave him useful answers, he cared little about the rigor of their demonstration. As he bluntly put it, "The best proof is to go and do it."[36]

An important disciplinary issue lay just beneath the surface of this conflict over standards of rigor, as Whittaker's remarks suggest. British mathematics was undergoing a major change of direction in the 1890s, particularly at Cambridge, and Heaviside had the misfortune of being caught in the resulting cross fire. Since Newton's time, Cambridge mathematics had been closely tied to physical applications, an emphasis that was institutionalized in the famous Tripos examination. The Mathematical Tripos was central to Cambridge education in the nineteenth century; it long provided the only route to an honors degree, and the private coaching of prospective "wranglers" was a substantial local industry. But the promotion of mathematics "for its own sake," or as a professional pursuit, was not the purpose of the Tripos; William Whewell, its chief architect at midcentury, said quite explicitly that it was designed "not to produce a school of eminent mathematicians, but to contribute to a Liberal Education of the highest kind."[37] Whewell regarded mathematics primarily as a means of mental training and a tool for the other sciences, and under his leadership the Mathematical Tripos emphasized intuitive clarity and problem-solving skills rather than abstraction or formal rigor. Cambridge turned out a long string of outstanding mathematical physicists between the 1840s and the 1870s, from Stokes and William Thomson to Maxwell, Rayleigh, and Joseph Larmor, but it remained notably—and deliberately—weak in pure mathematics and virtually oblivious to the work being done on the Continent.

35. Heaviside, *EMT* 2 (23 November 1894): 2.
36. Heaviside, *EMT* 2 (11 January 1895): 34.
37. Quoted in Harvey Becher, "William Whewell and Cambridge Mathematics," *Historical Studies in the Physical Sciences* 11 (1980): 1–48, on 32.

The high point in the development of the Mathematical Tripos came in the 1870s and early 1880s, when the nine-day examination was packed not only with the traditional dynamics, optics, and physical astronomy, but also with the newly added subjects of electricity, magnetism, and heat.[38] The competition for high places on the list of wranglers was intense, and as the examination became more and more demanding and elaborate, it eventually began to buckle under its own weight. Following the opening of the Cavendish Laboratory in 1874, physicists at Cambridge turned increasingly to experimental work, and in the late 1880s and 1890s J. J. Thomson led a series of reforms that resulted in the experimentally oriented Natural Sciences Tripos replacing the Mathematical Tripos as the main training ground for prospective physicists.[39] At the same time, an increasing number of young Cambridge mathematicians began to chafe under what they saw as the subordination of their subject to physics and other applications and sought to free themselves to pursue "pure mathematics" of the kind practiced in France and Germany. The work of A. R. Forsyth, E. W. Hobson, W. H. Young, G. H. Hardy—and William Burnside—gave fresh impetus to British mathematics between 1890 and 1910 and soon carried it far from its old traditions. Hobson and Forsyth led a long and finally successful campaign to reform the Mathematical Tripos itself, and within a few years after the turn of the century the orientation of the Cambridge school had shifted decisively from mathematical physics to pure mathematics.[40]

To secure both autonomy and support for their enterprise, the new generation of British pure mathematicians had to find a way to justify

38. For a portrait of this period, see A. R. Forsyth, "Old Tripos Days at Cambridge," *Mathematical Gazette* 19 (1938): 162–179.

39. David B. Wilson, "Experimentalists Among the Mathematicians: Physics in the Cambridge Natural Sciences Tripos, 1851–1900," *Historical Studies in the Physical Sciences* 12 (1982): 325–371.

40. I know of no satisfactory history of this transition, but it can be partially reconstructed from the obituaries and biographical notices of Forsyth, Hobson, Hardy, Young, and Burnside. Forsyth took the lead in introducing Continental pure mathematics to Cambridge in the early 1890s, but he soon fell behind. Leadership passed to Hardy and, a little later, J. E. Littlewood. On the reform of the Mathematical Tripos, see the text of the reform proposal (signed by Forsyth and Burnside, among others) in the *Cambridge University Reporter* 36 (1906): 740–746, and the subsequent discussion, 873–890. The mathematicians were concerned by their continuing loss of students to the Natural Sciences Tripos (enrollment in the Mathematical Tripos had dropped from 109 in 1890 to 56 in 1905) and sought to split the Mathematical Tripos into a general part that would attract large numbers of physics and engineering students and a specialized part intended for prospective professional mathematicians. On the "curious alteration in the popular subjects" that followed this reform, as advanced students moved away from applied mathematics and toward "subjects like the Theories of Functions and Groups," see W. W. Rouse Ball, "The Cambridge School of Mathematics," *Mathematical Gazette* 6 (July 1912): 311–323, on 323.

mathematical research as an end in itself, apart from any value it might have in training minds or in serving physics. If they were ever to win an independent professional status for themselves and their discipline, they had to break the hold of the belief, dominant at Cambridge since Whewell's time, that mathematical studies were primarily a means to other ends. They did so by drawing on the ideology of "pure mathematics" that had long flourished on the Continent, which asserted that the true value of mathematics lay not in its fruits, but in the permanence, beauty, and rigor of its results. C. G. J. Jacobi gave a classic statement of this Idealist view in 1830 in response to Fourier's claim that the principal object of mathematics was the service of human needs and the explanation of natural phenomena. "A philosopher like him ought to know," Jacobi declared, "that the sole object of the science is the honor of the human spirit and under this view a problem of [the theory of] numbers is worth as much as a problem on the system of the world."[41] G. H. Hardy went even further in his 1940 *Mathematician's Apology*, rejoicing that it was the "very remoteness from ordinary human activities" of pure mathematics that helped to "keep it gentle and clean."[42] Far from being a reproach, the "uselessness" of pure mathematics became a point in its favor. Acceptance of this view served to raise pure mathematics above other scientific disciplines; instead of being a mere tool, it became the master science, unsullied by mundane concerns and rising highest toward absolute truth. It was a view pure mathematicians found very appealing, of course, for it placed a high value on their work while freeing them to follow their curiosity wherever it might lead.

The ideology of pure mathematics was reflected in the rhetorical practices of its adherents, which differed in important ways from those of natural scientists. Like other scientists, pure mathematicians sought to make their statements "strong" by invoking the support of respected authorities, by stressing the generality and formal beauty of their results, and by linking their claims to accepted bodies of knowledge.[43] But where

---

41. C. G. J. Jacobi to A.-M. Legendre, 2 July 1830, quoted in Kline (n. 8), 813; cf. Morris Kline, *Mathematics: The Loss of Certainty* (New York: Oxford University Press, 1980), 286.

42. G. H. Hardy, *A Mathematician's Apology* (Cambridge: Cambridge University Press, 1940; 2d ed., 1967), 121. Hardy sought to defend mathematics mainly as a "creative art," and so was not so insistent on strict rigor as were some of his colleagues, although he agreed that rigorous proofs eventually had to be supplied.

43. The transformation of scientific statements into "stronger" types is discussed in Bruno Latour and Steve Woolgar, *Laboratory Life: The Construction of Scientific Facts* (Princeton, N.J.: Princeton University Press, 1986), 81–86, and in Bruno Latour, *Science in Action* (Cambridge, Mass.: Harvard University Press, 1987), 21–62, but without specific attention to pure mathematics.

scientists in other disciplines made such links in several ways, mathematicians recognized only one type of link—that of pure deduction—as carrying full persuasive force. Natural scientists placed a high value on empirical evidence and regarded the display of a close fit with observations as an especially effective and convincing rhetorical move. Engineers, of course, leaned even more strongly toward the view that the best argument for the truth of a proposition was to show that it worked in practice. But the ideology of pure mathematics left no room for such appeals to experience. In the words of Gottlob Frege, perhaps the strictest of the rigorists, "If we ask what gives mathematical knowledge its value, then the answer must be that the value consists less in *what* is known than *how* it is known, less in the content of the knowledge than in the degree of the self-evidence and the insight into logical connections."[44] Having staked the justification for their enterprise in part on its claim to absolute rigor, pure mathematicians could not admit the validity in mathematics of any form of argument that was not fully rigorous. They had taken their stand on the circumscribed but seemingly secure ground of pure deduction, and the argument that a technique or result "worked" or met some empirical test could thus carry no persuasive force in mathematical discourse; it did not have the power, as in the natural sciences, to move a statement along the path from "conjecture" to "fact." The only way to make a statement truly strong in pure mathematics was to demonstrate the rigor of its derivation. Conversely, the best way to weaken an opponent's argument was to point out deficiencies in its rigor. Even a result that had been abundantly confirmed in other ways was, as Burnside said, "absolutely valueless" to pure mathematicians if it lacked a strictly logical derivation.

Burnside's purpose in 1894 was quite clear: to keep Heaviside's unrigorous writings out of the *Proceedings of the Royal Society*. But his actions can be fully understood only when placed in their broader historical and disciplinary context. Heaviside's paper reflected the standards and approach of "the happy old easy-going Todhunter period" of mid-nineteenth-century Cambridge, when problem-solving and physical applications counted for everything and abstract rigor for almost nothing.[45] It embodied—indeed,

<hr/>

44. Quoted in Philip Kitcher, "Mathematical Rigor—Who Needs It?" *Nous* 15 (1981): 469–493, on 470. After a very perceptive analysis of the uses of rigorization, Kitcher concludes (p. 490) that mathematicians' claims that their work is driven by a desire for absolute rigor are merely "misguided window-dressing." My question is: why *this* window dressing? What function have claims about rigor served for the mathematicians who have made them?

45. Cooper (n. 6), 13, says that Heaviside's work was "typical of the British mathematics of the 1850s," while Lützen (n. 6), 175n, suggests that he worked within "an otherwise abandoned paradigm from about 1800."

in an exaggerated form—everything Burnside and his allies were fighting against. As pure mathematics pulled away from physics in Britain in the 1890s, the new proponents of rigor moved to assert control over the standards in their field and to clear a boundary zone around their emerging discipline. If physicists or electrical engineers wanted to devise tricks to help them solve circuit problems, that was fine, even if they sometimes used such absurdities as fractional differentiation and divergent series. But if they tried to present such tricks as contributions to serious mathematics, the rigorists had to draw the line or risk undermining the foundations of their own enterprise. By 1894 there was no longer room in mathematics for a paper like Heaviside's.

Aftermath

By the rules of the Royal Society, papers submitted to it but found unsuitable for publication were not returned to their authors, but were instead deposited permanently in its Archives. This would have been the fate of Heaviside's paper, but shortly after Rayleigh informed him in July 1894 that his paper had been rejected, it was discovered that, through an oversight, Part III had never been formally "read" at a meeting, and he was offered an opportunity to withdraw it. Heaviside replied tartly that he would have been "better pleased to have heard that the oversight was of another nature, and that it was determined to print the paper in due course; as it is of course obviously suggested that it is not worth printing; an idea which cannot be entertained for a moment by the Author thereof!" He added, however, that if the decision not to accept his paper was irrevocable, "I should, with much reluctance, prefer to withdraw it, in order to seek another way of publishing it, which will be more effective than the Archives."[46]

Once he got his paper back, Heaviside spent several months going over it—"to see if I could make nonsense of it," he told FitzGerald early in 1895. "Also, like D. Swiveller, waiting for something to turn up; quite indefinite the something."[47] He concluded that the paper was not nonsense, but his injured pride made him slow to submit it elsewhere. His friend John Perry, a London engineering professor and a great advocate of the operational

46. Rayleigh to Heaviside, 26 July 1894, OH-IEE; Heaviside to "Secretary of the Royal Society," 5 September 1894, RR.12.136, RS Archives. It is unclear whether this was a genuine oversight or whether Rayleigh bent the rules in order to return the paper to Heaviside.

47. Heaviside to FitzGerald, 4 February 1895, FG-RDS. "Dick Swiveller" was a character in Dickens's *Old Curiosity Shop*.

calculus, urged him to publish Part III in the *Proceedings of the Physical Society*, which would not object to any supposed lack of rigor.[48] But Heaviside resisted; his paper was a contribution to mathematics, not physics, he said, and should be published as such. He also turned down an offer from FitzGerald to try to publish it in the *Transactions of the Royal Irish Academy.* "The Royal Society is the proper place, if anywhere," he insisted; Part III deserved to be published alongside its predecessors, not taken on sufferance by some other journal.[49] Finally, and somewhat reluctantly, he published a "boiled down" version in his regular series of articles in the *Electrician* late in 1898 (reprinted the next year in Volume II of his *Electromagnetic Theory*), but this did little to soften his bitterness.[50]

Heaviside regarded the rejection of his paper as a direct insult, and he broke most of his ties with the Royal Society after 1894. FitzGerald dissuaded him from resigning in protest or taking some other "extreme step," but Heaviside never again submitted a paper to the Society and twice pointedly refused to accept its Hughes Medal, even when he was in dire need of the £100 that went with it.[51] It seemed, he said, that the Society existed "not merely for the encouragement of research along established lines, but also for the active discouragement of work of a less conventional character," and he had little desire to be associated with such an obstructive and "unpractical" body.[52]

Heaviside's real target, however, was not the Royal Society itself, but the "rigorists" within it—"mathematicians of the Cambridge or conservatory kind," as he called them, "who look the gift-horse in the mouth and shake their heads with solemn smile, or go from Dan to Beersheba and say that all is barren."[53] The "academical system of rigorous mathematical training" had its merits, he said, but it also had a serious fault: "it checks instead of stimulating any originality the student may possess, by keeping him in regular grooves." Thus it made those who passed through it unable to ap-

48. Heaviside to FitzGerald, 4 February 1895, FG-RDS. Perry was then using Heaviside's methods to solve heat flow problems in order to bring out flaws in Kelvin's theories about the age of the earth; see Nahin (n. 1), 245–256.

49. Heaviside to FitzGerald, 4 February 1895, FG-RDS; cf. FitzGerald to Heaviside, 8 February 1895, OH-IEE.

50. Heaviside, *EMT* 2 (7 October, 2 December 1898): 457–482.

51. FitzGerald to Heaviside, 20 August 1894, OH-IEE. This is a reply to an earlier letter from Heaviside that has not survived. On Heaviside's refusal of the Hughes Medal, see Joseph Larmor to Heaviside, 3 November 1904 and 10 July 1908,OH-IEE, and Heaviside to Larmor, 4 November 1904 and 15 July 1908, Larmor Papers, RS Archives.

52. Heaviside, *EMT* 2 (23 November 1894): 3–4; Heaviside to FitzGerald, 5 August 1894, FG-RDS.

53. Heaviside, *EMT* 2 (14 December 1894): 12.

preciate work—notably Heaviside's—that lay outside the usual grooves.[54] Worse, it sometimes made them actively hostile to such work. It was simply narrow-minded, Heaviside said, for a referee to reject the operational calculus as unrigorous unless he could put something better in its place, or at least explain how such an objectionable method could so often give correct answers. Writing to FitzGerald a few months before his own paper was rejected, Heaviside said that a referee ought to approve even a paper that contained serious errors if it also included "a substantial amount of good." "What does it matter if there is some nonsense in a paper?" he asked. "It will die, and the good will live."[55] Time should do the sifting, he said, not over-scrupulous referees.

The rejection of his paper stung Heaviside into attacking the rigorists' assumptions on a whole range of issues, from the nature and origin of mathematical knowledge to the proper place of rigor in mathematical discourse. Urged on by John Perry, he began to express his views publicly in a series of articles in the *Electrician* in the fall of 1894. In these articles, and in remarks scattered through his later writings, Heaviside presented a sharply drawn critique of the rigorists' insistence on a purely deductive mathematics and a defense of a more open and flexible approach modeled on that of the natural sciences.

Heaviside's starting point was his conviction that mathematics is an *experimental* science. All of its branches, he said, had developed by a process of abstraction and idealization from our experience of the physical world. Geometry, for example, was simply a codification of our experience of space, and the belief common among mathematicians that its truths were somehow "pre-existent in the human mind" was, Heaviside said, badly mistaken:

> You might assert the same pre-existence of dynamics or chemistry. I think it is a complete reversal of the natural order of ideas. It seems to me that geometry is only pre-existent in this limited sense: that since we are the children of many fathers and mothers, all of whom grew up and developed their minds (so far as they went) in contact with nature, of which they were a part, so our brains have grown to suit. So the child takes in the facts of space geometry naturally and easily. The experience of past generations makes the acquisition

54. Heaviside, *EMT* 2 (11 January 1895): 32.
55. Heaviside to FitzGerald, 13 April 1894, FG-RDS. These remarks concerned a paper by Joseph Larmor that FitzGerald was refereeing for the *Philosophical Transactions*; although Heaviside thought the paper was seriously flawed, he advised FitzGerald to recommend its acceptance.

of present experience easier, and so it comes about that we cannot help seeing it. But it is all experience, after all; although learned philosophers, by long, long thinking over the theory of groups and other abstruse high developments, may perhaps come to what I think is a sort of self-deception, and think that their geometry is pre-existent in themselves, and nature's is only a bad copy.[56]

Mathematicians' attempts to deny the experiential foundations of their subject were just a sham, Heaviside said; although the real work of mathematical discovery was "largely experimental," and proceeded mainly by tentative generalization and trial and error, "rigorous mathematicians conceal this as well as they can when they write treatises, and pretend to be omniscient by avoiding their failures, and also the many interesting things they cannot explain."[57] In an effort to preserve the illusion that they advanced only by the sure steps of deduction—a key element of their argument for the value and autonomy of their discipline—mathematicians pruned and rearranged their findings in such a way that, once done, they could "put the logic at the beginning, and pretend they knew all about it before they began."[58] By shifting into a context in which their own rhetorical tools—those of pure deduction—were stronger, mathematicians were able to make it appear that their knowledge descended from above, from pure logic, rather than rising from the ground of experience.

Heaviside did not mean to denigrate logic, which he regarded as a necessary tool for ordering and clarifying our ideas. "But there is logic *and* logic," he said, and he had little patience with the "narrow-minded logic confined within narrow limits, rather conceited, and professing to be very exact" that the rigorists insisted upon. He preferred "a broader sort of logic, more common-sensical, wider in its premisses, with less pretension to exactness, and more allowance for human error, and more room for growth."[59] He particularly objected to the widespread belief that strict logic and explicit definitions had to come first in mathematics, "or else you prove nothing."[60] Reliable mathematical knowledge could be, and gener-

56. Heaviside, "The Teaching of Mathematics," *Nature* 62 (4 October 1900): 548–549; reprint Heaviside, *EMT* 3: 513. It is tempting to take the remark about groups as a jab at Burnside, but there is no evidence that Heaviside knew of Burnside's role in blocking his paper. There are obvious affinities between Heaviside's empiricist philosophy of mathematics and J. S. Mill's, but Heaviside apparently developed his ideas independently.
57. Heaviside, *EMT* 3: 234.
58. Heaviside, *EMT* 3: 370. Scientists (and even historians) do something very similar in writing up their findings, of course, but the key point is that mathematicians generally deny that they start from an empirical base.
59. Heaviside, *EMT* 3: 516–517.
60. Heaviside, *EMT* 3: 370.

ally was, acquired in a more empirical and exploratory way, he said, just as scientific knowledge was, and complaints about a lack of rigor in the derivation of a procedure that had proven itself in practice were beside the point. "Shall I refuse my dinner because I do not fully understand the process of digestion?" he asked. "No, not if I am satisfied with the result."[61] The proper place for logic and rigor was at the end, as a codification of truths already known. Consider geometry, not as it is presented in Euclid, but as it is actually done and learned: "A straight line can never be intelligibly defined *per se*," Heaviside said. "One must actually know the practical straight line before any definition of the abstract straight line can be understood. Then our understanding and acceptance of the definition is a recognition that it states what we knew already, in accumulated experience, though we may have never openly thought about it." Continuing in a very modern-sounding vein, he asserted that even such a "final" codification was necessarily incomplete:

> There is also no self-contained theory possible, even of geometry considered merely as a logical science, apart from practical meaning. For a language is used in its enunciation, which implies that developed ideas and complicated processes of thought are already in existence, besides the general experience associated therewith. We define a thing in a phrase, using words. These words have to be explained in other words, and so on, for ever, in a complicated maze. There is no bottom to anything. We are all antipodeans and upside down.[62]

Heaviside had little sympathy with the search for a purely formal or logical foundation for mathematics, and so found himself out of step with the movement toward "absolute rigor" that absorbed the energies of so many leading philosophers and mathematicians around the turn of the century.[63] It was a doomed effort, he thought, and an empty and illusory goal.

Among the many scientists and engineers who shared Heaviside's views about the rigorists was Rayleigh. Rayleigh later expressed regret that his duties as Secretary of the Royal Society had forced him to play a part in the mathematicians' rejection of Heaviside's paper, and it was no doubt with this episode in mind that he inserted a paragraph on "rigour" into the preface of the second edition of his *Theory of Sound*, completed in July 1894. "In the mathematical investigations," he wrote,

61. Heaviside, *EMT* 2 (23 November 1894): 9.
62. Heaviside, *EMT* 2 (23 November 1894): 2–3.
63. The rise and fall of the search for rigorous foundations is traced in Kline (n. 8), 172–306.

I have usually employed such methods as present themselves naturally to a physicist. The pure mathematician will complain, and (it must be confessed) sometimes with justice, of deficient rigour. But to this question there are two sides. For, however important it may be to maintain a uniformly high standard in pure mathematics, the physicist may occasionally do well to rest content with arguments which are fairly satisfactory and conclusive from his point of view. To his mind, exercised in a different order of ideas, the more severe procedure of the pure mathematician may appear not more but less demonstrative.[64]

Heaviside quoted this passage "with much pleasure" in the *Electrician* in November 1894, and drew particular attention to the words "not more but less demonstrative."[65] His point, like Rayleigh's, was simple but important: standards of argument and evidence are not absolute and universal. Arguments that are found fully convincing by those in one discipline may seem "less demonstrative" to those in another; in particular, the "rigorous" reasoning of a pure mathematician may appear less conclusive to a physicist than an argument founded on physical considerations or empirical evidence.

Ironically, Burnside had been making a similar point, although from the other side: in rejecting Heaviside's paper, he had said, in effect, that arguments that seemed quite convincing to a physicist could fail completely to meet the standards of pure mathematicians. Each discipline had its own aims and interests and its own corresponding set of accepted rhetorical practices. Arguments not framed in accordance with the accepted practices of a discipline could be seen as a threat to its integrity, as Heaviside's "unrigorous" mathematics was in the eyes of pure mathematicians. To protect their interests, and ultimately their autonomy, mathematicians felt they had to take "disciplinary" action, as Heaviside discovered in 1894.

Heaviside's operational calculus and most of his results on fractional differentiation and divergent series were "rigorized" by the 1930s and have now passed into the body of accepted mathematics, though often in the guise of variants on much older work.[66] As E. T. Bell later described the

64. Rayleigh, *The Theory of Sound,* 2d ed. (London: Macmillan, 1894; reprint New York: Dover, 1945), xxxv.
65. Heaviside, *EMT* 2 (23 November 1894): 5; cf. Rayleigh to Heaviside, 24 October 1912, OH-IEE, in which Rayleigh says that "though educated at Cambridge I am largely with you in repudiating the claims of the 'rigourists,' and I agree that the mere logic should come last and not first."
66. Lützen (n. 6).

process, "Following the trite pattern, the Heaviside tragi-comedy degenerated in three acts into broad farce: the Heaviside method was utter nonsense; it was right, and could be readily justified; everybody had known all about it long before Heaviside used it, and it was in fact almost a trivial commonplace of classical analysis."[67] Once again, mathematicians had managed to "put the logic at the beginning, and pretend that they knew all about it before they began." By casting Heaviside's results into a new idiom in which they appeared as deductive rather than empirical truths, mathematicians were able to defuse the threat posed by his work and to reinforce their claim to special status as a uniquely rigorous discipline.

I would like to thank the Royal Society of London, the Institution of Electrical Engineers, the Royal Dublin Society, University College London, and the American Institute of Physics for permission to quote materials in their possession.

67. E. T. Bell, *The Development of Mathematics* (New York: McGraw-Hill, 1940), 389.

Part II

Discipline and Epistemology

Lissa Roberts

4. Setting the Table: The Disciplinary Development of Eighteenth-Century Chemistry as Read Through the Changing Structure of Its Tables

Introduction

ANTOINE LAVOISIER introduced his epochal textbook, the *Traité élémentaire de chymie*, with the following claim.

> . . . while I thought myself employed only in forming a Nomenclature . . . my work transformed itself by degrees, without my being able to prevent it, into a treatise upon the Elements of Chemistry.[1]

The genesis and development of this essay can be described in similar terms. My own work in the history of chemistry began as an idealist study of language. By contrasting Priestley's and Lavoisier's choices of nomenclature and rhetoric of reporting, I hoped to learn something about the nature of the chemical revolution. But, as I proceeded, I found myself confronting the inextricable link between language and constructive practice in the formation of chemical knowledge.

Indeed, I found generally that beneath the apparently placid surface of eighteenth-century chemistry's history of linguistic organization and reform—traditionally depicted as a linear march of progress[2]—rests a deeper history that reflectively portrays chemistry's disciplinary journey from its self-defined status as art to its recognized status as science. In other words, eighteenth-century attempts to discipline chemistry linguistically were

---

1. Antoine Laurent Lavoisier, *Traité élémentaire de chymie* (Paris, 1789). All quotations taken from *Elements of Chemistry,* translated by Robert Kerr, with introduction by Douglas McKie (New York: Dover, 1965, originally published in Edinburgh, 1790). See p. xiv.
2. The most exemplary purveyor of this view is Maurice Crosland, *Historical Studies in the Language of Chemistry* (New York: Dover, 1978).

pregnant with disciplinary attempts (understood in the broader, praxical sense) to establish chemistry as a science. And, when we definitionally utilize the ideal structures and goals inherent in what eighteenth-century chemical practitioners took to be "art" and "science," we find that nomenclatural and linguistic reform—represented in this article by the structure and content of chemical tables—embodied the revolutionary redirection that Lavoisier's chemistry lent to the discipline writ large.

The best recent exposition of how language and practice (both laboratory and social) have inseparably forged scientific knowledge can be found in Steven Shapin's article, "Pump and Circumstance: Robert Boyle's Literary Technology."[3] In his article, Shapin delineates three types of technology which, since the time of Robert Boyle, have been interdependently employed in the construction and social acceptance of science: material, social, and literary technologies. While viewing all three as mutually embedded, manipulative instruments of investigation and persuasion, he specifically focuses on literary technology ("the expository means by which matters of fact were established and assent mobilized") to argue against the received notion that production and communication of knowledge are distinct endeavors.[4]

Shapin writes that "speech about natural reality is a means of generating knowledge about reality, of securing assent to that knowledge and of bounding domains of certain knowledge from areas of less certain standing."[5] In his own work, he analyzes literary technology primarily in terms of its deployment by actors; that is, in terms of authors' strategies of exposition and persuasion. In this essay, I will instead analyze a series of chemical tables (prime examples of eighteenth-century chemistry's literary technology) to expose the underlying structures that informed their modes of exposition, which, together with the changing structures of chemistry's other technologies, provided the particularly structured space of possibilities wherein knowledge of nature was generated, organized, and articulated.

This essay, then, is only one step toward establishing a historical method capable of providing the richest possible interpretation of primary sources. By wedding sensitivity toward the structuring impact of language at all levels of exposition (as well as that of instrumentation and social institutions) to the sort of in-depth archival research practiced by historians such

3. Steven Shapin, "Pump and Circumstance," *Social Studies of Science* 14 (1984): 481–520. See also Shapin and Simon Schaffer, *Leviathan and the Air-Pump* (Princeton, N.J.: Princeton University Press, 1985).

4. Ibid., 484.

5. Ibid., 481.

as F. L. Holmes,[6] we could construct a richly nuanced presentation of eighteenth-century chemistry. For now, the scope of this essay is more modest. Primarily through an analysis of exemplary chemical tables, it presents the history of eighteenth-century chemistry's disciplinary development as involving a movement from the depiction and practice of chemistry as a manipulative art in the perceived context of transcendent nature to its establishment as a science, structured in such a way as to identify nature with both the process and product of chemical manipulation.

As chemistry began to emerge publicly as an autonomous discipline in the first decades of the eighteenth century, it did so as a self-proclaimed art. Defined and circumscribed by a set of laboratory techniques, chemistry made no claim to possessing a fully systematic understanding of nature. That is, it artfully opened the world of nature to scrutiny and manipulation without prescribing any single interpretive structure as necessary for nature's full exposition and understanding. From the 1750s, chemists increasingly lobbied for the acceptance of their discipline as a science uniquely grounded in a set of distinctly chemical principles and practices. In this process, chemistry's artisanal heritage was institutionalized as a disciplinary precept; the chemical community was publicly bound by the rhetoric of theoretical neutrality. Chemists held that through experimentation and a cooperative accrual of matters of fact—unhindered by disputes over those facts' systematically theoretical significance—they might ultimately induce nature to reveal the secrets of its composition and organization.

The coming of the chemical revolution radically altered this situation. The science of chemistry was now to begin with a set of principles that predetermined the structure of nature even as chemists set out to investigate it. Once "naturalized" as the primary elements oxygen and caloric, chemistry's manipulative reach stretched out to encompass and order the entire world of inquiry. Individual discoveries had still to be made, but their construal and position in the general order of things were already mandated by the language and practice in which they were formed.

In this context, the significance of chemical tables is that they embodied certain categories and relations with which to structure the world of investigation, while disallowing others.[7] As will be seen, this was true both in

6. See F. L. Holmes, *Lavoisier and the Chemistry of Life* (Madison: University of Wisconsin Press, 1985). For a general overview of the current state of research on the chemical revolution, see Arthur Donovan, ed., *Osiris* 4 (1988), which is dedicated to a reassessment of the field.

7. In this sense, the paradoxical history of technical art is laid bare. As art, its products and direction were always already distanced from nature and subject to human construction. In Heideggarian terminology, it is only a question of historically witnessing the "coming to presence" of technology's manipulative dominance in the establishment of modern science.

terms of their presentational structure and the nomenclature they employed. For most of the century, chemical tables recorded individual findings of affinity or solvent-solute relations in a post-facto manner, without predetermining either the outcome of other individual relations or how nature itself was structured. Chemists ranging from E. F. Geoffroy in 1718 to Torbern Bergman in the early 1780s insisted that their tabulated results not be taken a priori as generalizable statements about nature. One had first (at least) to examine "all possible combinations to assure that there is nothing to the contrary."[8]

In a fairly sudden transition during the late 1780s, chemistry came to be informed by a wholly new sort of table. Lavoisier's *Traité élémentaire* abounds with tables that embody highly specific categories based on processes, such as oxygenation, which spread out to organize the investigative understanding of nature through a set of predetermined research projects. These processes were themselves "naturalized" by their reified embodiment in the list of primary elements that headed Lavoisier's instrumentally determined "table of simple substances." In place of the once transcendent "chain of nature" revealed and recorded link by link in "prerevolutionary" synoptic tables, chemistry's tables (and the instrumental practices embedded in them) gained a powerful priority by the end of the century to fashion nature anew.

Chemistry's Art Bears a Table

In the first decades of the eighteenth century, chemistry had not yet attained the status of a distinct scientific discipline. Lacking much of a history as public (as opposed to occult) practice, an established set of organizing principles and autonomous institutional standing,[9] it was largely accepted as an art in the service of physics, physick, industry (dyeing, metallurgy, and

---

See Martin Heidegger, "The Question Conerning Technology," *The Question Concerning Technology and Other Essays* (New York: Harper and Row, 1977), 3–35.

8. *Mémoires de l'académie royale des sciences* (Paris, 1718), 202–212. All quotations (modified to indicate problems of translation) taken from H. M. Leicester and H. S. Klickstein, eds., *A Sourcebook in Chemistry 1400–1900* (Cambridge, Mass.: Cambridge University Press, 1968), 67–75, on p. 68. Cited as "Table."

9. Robert Boyle's contributions might be viewed as the commencement of that history. In terms of chemistry's institutional standing, consider that the first chairs of chemistry at Cambridge and Oxford were not established until the first decade of the eighteenth century. Chairs in Sweden—a major center of chemical activity—were not established until the 1750s–1760s.

so forth), and, finally, natural theology. Its craft status was manifest in any number of contemporary definitions of chemistry. Struggling against identification with the esoteric tenor of alchemy, Richard Russell defined it as early as 1678 in the following terms. "Chemistry is a true and real Art, and (when handled by prudent Artists) produceth true and real effects."[10] John Freind (1675–1728), Oxford University's first professor of chemistry (in 1704), set the tone for the first three or four decades of the eighteenth century by limiting chemistry's reach to the sensible laboratory practices of dissolution and combination. He wrote that "chemistry is the art of conjoining separate parts of natural bodies and of dividing them when conjoined."[11]

As late as 1735, Herman Boerhaave (1668–1738) maintained this definitional tone in his highly influential elementary textbook. He wrote:

Chemistry is an art, that teaches us how to perform certain physical operations, by which bodies that are discernible by the senses, or that may be rendered so, and that are capable of being contained in vessels, may by suitable instruments be so changed, that particular determined effects may be thence produced, and the causes of those effects understood by the effects themselves, to the manifold improvement of various arts.[12]

According to such definitions, chemistry encompassed only the work of the hand and the world of the senses, directed toward explicitly utilitarian ends.[13]

It was in this context that, in 1718, Etienne François Geoffroy (1672–1731) published his "Table des différens rapports observés entre différentes substances," initiating a new genre of experimental reporting in chemistry. With this table Geoffroy provided a handy compendium listing the relative attraction of a variety of chemical substances for each other. Across the top

10. This definition is drawn from the translator's preface to the reader in *The Works of Geber Englished by Richard Russell*, quoted in Maurice Crosland, "Chemistry and the Chemical Revolution," G. S. Rousseau and Roy Porter, eds., in *Ferment of Knowledge* (Cambridge: Cambridge University Press, 1980), 389–416, on 389.

11. John Freind, *Chemical Lectures: In Which Almost All the Operations of Chemistry Are Reduced to Their True Principles, and the Laws of Nature*, trans. Philip Gwillim (London, 1712) 10. As indicated by the disjuncture between Freind's definition of chemistry and the title of his book, "true principles" had to be drawn from outside the practice of chemistry proper.

12. Hermann Boerhaave, *Elements of Chemistry*, trans. Timothy Dallowe M.D. (London, 1735),19.

13. Along these lines, it is interesting to note that in his *Encyclopédie* article "Chymie" (discussed later in this essay), Venel described Boerhaave's work on fire as consisting of a few, initial chemical observations followed by the bulk of the treatise which offered physical explanations. Diderot and d'Alembert, eds., *Encyclopédie* (Paris, 1753), 3: 408–437.

row he placed sixteen chemical reagents—or types of reagents—below which he arranged different substances and types of substances in descending order of their attraction to the substance at the head of their column.

The first thing to note regarding this table is its title. Geoffroy described the relations he recorded as *rapports*. While usually translated into English as "affinities," the term he chose denoted no particular causal implication; it signified only that a relation had been observed.[14] By selecting this term, Geoffroy distinguished between the "natural" causal act responsible for chemical combination and the effects that he and his fellow chemists observed, manipulated, and recorded. This was not because he had no ideas about what caused the chemical phenomena he effected and observed. It was not because he was some kind of positivist *avant la lettre*. In his *Treatise of the Fossil, Vegetable, and Animal Substances that are made Use of in Physick*[15] Geoffroy was quite explicit in his explanations of material combination. But the disciplinary context in which he wrote here was quite different. A philosophical treatise on medicine allowed for, indeed assumed, causal explanation. A memoir on chemistry, however, apparently did not.

In the *Treatise*, Geoffroy described all bodies as compounds of three elementary principles: fire (the primary cause of all motion), water, and earth. He considered a compound's particular properties to result from the specific combination of these principles, which he explained in terms of particulate shape and activity. Fire was subtle and active, easily permeating the pores of all other bodies, setting them in motion and giving rise to heat. Water was "conceived to consist of small smooth particles of an oblong or oval figure, and perfectly rigid or inflexible." This accounted for its great penetrability and equal ease of separation. "The want of taste or smell in water seems to be owing to the smoothness, obtuseness, and smallness of its particles, which cannot vellicate the nerves of the tongue or nostrils." So too was elementary shape and combinative ease with fire responsible for water's fluidity and transparence. Earth particles were irregularly shaped and most ponderous. As such, they gave bodies solidity and, depending on their combination with fire and water, accounted for more specific properties owing to the shape they lent compound bodies. Acid particles, for example, had sharp points.[16]

14. Geoffroy, "Table." Failure to note such subtleties in language blocks the historian's entry into the nuances of science's historical development.

15. E. F. Geoffroy, *A Treatise of the Fossil, Vegetable, and Animal Substances* (London, 1736).

16. Geoffroy, *Treatise*, 8–12. For a discussion of Geoffroy's theoretical views, see W. A. Smeaton, "E. F. Geoffroy Was Not a Newtonian Chemist" (letter to the editor) *Ambix* 18 (1971): 212–214.

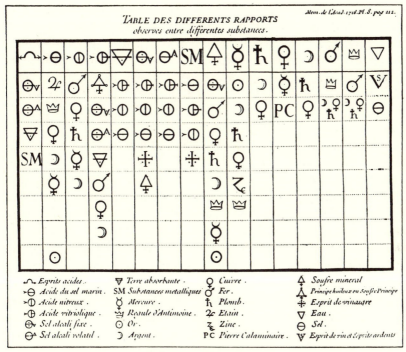

Geoffroy's table of affinities.

Plate I. Etienne François Geoffroy, "Table des différens rapports observés entre différentes substances." *Mémoires de l'académie des sciences* (Paris, 1718), 202–212.

However, nowhere in Geoffroy's table or the memoir in which it appeared was the question of substances' elemental or causal properties broached. Although at one point Geoffroy alluded to "a very subtle sulfurous principle" as possibly responsible for a particular solvent reaction, he added quickly that "This is not the place to examine this subject."[17] Rather

17. Geoffroy, "Table," 73. This presentation of Geoffroy differs from that of Hélène Metzger who attributes Geoffroy's success to the contemporary prestige of Stahl's phlogiston theory. Note that phlogiston does not even appear as an element in Geoffroy's table. Given Metzger's own recognition of the centrality of experimentation in the development and deployment of affinity tables specifically and chemistry in general, we have even more cause to question the validity of a theory-dominant approach to the history of chemistry. In retrospect, Lavoisier did compare Geoffroy's sulphur to Stahl's phlogiston, but this is hardly relevant evidence. See Hélène Metzger, *Les doctrines chimiques en France du début de XVIIe à la fin du XVIIIe siècle* (Paris: Albert Blanchard, 1969), 405, 412–413.

than direct chemists toward a search for primary components and causes, his table simply recorded the perceived relations among (admittedly compound) substances with which chemists worked in their laboratories. Since manipulative familiarity and efficacy rather than exact knowledge were the goal of chemistry's art, quantitative exactitude and substantial purity were unimportant. Indeed, Geoffroy ended his memoir with the comment that "in many of these experiments the separation of materials is not perfectly exact and precise."[18]

It was this limited criterion of utility which marked the selection of chemical substances and their relations recorded in Geoffroy's table and generally situated the construction of such tables in relation to the attainment of natural knowledge. Geoffroy's selection ranged from the first column of the table in which he compared "acid spirits" with other general types of bodies (fixed alkali salts, volatile alkali salts, absorbent earths, and metallic substances), to the more specific relations charted in columns 9 through 16 where he compared individual (but nonetheless, compound) substances to a variety of other individual substances. The structure and content of his table thus directed chemists away from investigating underlying causes, leading them instead to a general proposition describing all such observable relations. But even here, Geoffroy the chemist[19] provided little certainty, commenting that "one cannot give it as general without having examined all possible combinations to assure that there is nothing to the contrary."[20]

> The proposition itself, then, was but an inductively obtained description. It states: Whenever two substances which have some disposition to unite, the one with the other, are united together and a third which has more *rapport* for one of the two is added, the third will unite with one of these, separating it from the other.[21]

As such, this proposition left chemists in the laboratory to continue amassing and organizing observed relations. For now, principles derived from some other, already established discipline(s) were required to give chemis-

For a trenchant critique of Metzger's historiography, see John Christie, "Narrative and Rhetoric in Hélène Metzger's Historiography of Eighteenth-Century Chemistry," *History of Science* 25 (1987): 99–109.

18. Geoffroy, "Table," 74–75.

19. Geoffroy is referred to here as a chemist to distinguish his role as chemical practitioner and commentator from other aspects of his work. Depending on the context and discourse under question, the same actor might be considered in any number of different ways.

20. Geoffroy, "Table," 68.

21. Ibid., 68.

try greater purpose and/or reveal what was considered the actual structure of nature.

When chemists did come to offer distinctly chemical principles of disciplinary organization and knowledge—that is, principles not drawn from outside the domain of chemistry's practice—they were faced with nothing less than having to transmute chemistry from an art into a science. Peter Shaw (1694–1763), important as a chemist and lecturer in his own right as well as the English translator of seminal works by Georg Ernst Stahl (1660–1734) and Hermann Boerhaave, recommended that chemistry be disciplined by digesting experiments "into tables; showing what they prove, and how far they reach."[22] More specifically, both he and Boerhaave advocated the construction of tables in which to record the solvent activity of menstruums. In Boerhaave's words,

> If it was possible, now, to accomplish, that *menstruums* should be reduced into order, according to the difference of their manner of acting, and then be distributed into inferior classes, then the doctrine of chemistry might be brought to the certainty of a science."[23]

While focusing on different operations (dissolution rather than combination), these men sought to discipline chemistry through reliance on the same sort of synoptic tables by which Geoffroy had offered simply to advance chemical practice. Rather than relate laboratory findings to some primary causal principle(s) like so many points on or in a sphere to its center, these chemists sought to redirect the role of synoptic tables from useful repositories of experimental results to the means whereby chemical practice would somehow be placed on a scientific footing. But, as we shall see, the structure of chemistry tables, though much embroidered, remained essentially the same until the late 1780s. Chemists hoped that by increasing their manipulative grasp on nature's chemical effects and cataloging the results, they would literally induce nature to bare the secrets of its own structure.

Does Practice Make Perfect?

The history of chemistry which followed in the period leading up to its self-proclaimed revolution (from the 1750s through the 1780s) was marked by

22. Peter Shaw, *Chemical Lectures* (London, 1734), 436.
23. Hermann Boerhaave, *Elements of Chemistry*, 398.

the proliferation, growing complexity, and sophistication of its synoptic tables.[24] But the increasingly intricate set of facts that chemists produced and recorded did not yield the natural revelation they continued to hope for. By maintaining and championing a rhetorical stance of theoretical neutrality throughout the period—all the while tacitly informing their work through the direction, tools, and language of their investigation— chemists ultimately succeeded in inducing interpretive conflict rather than a systematic understanding of nature. Under the banner of "revolutionary" chemistry, a wholly new system of organization would finally fill the breach created by this conflict, a system that powerfully asserted nature's structure from the outset. In the interim, though, chemists worked to discipline themselves as much as their subject matter, laying down guidelines for proper behavior alongside their growing fund of investigative results.

Their efforts in this direction can be seen most overtly, perhaps, in the methodological prescripts and general discourse of contemporary chemistry texts. As documented by Wilda Anderson, Pierre Joseph Macquer (1718–1784) prefaced and postfaced his widely popular *Dictionnaire de chymie* by urging the establishment of a cooperative community of natural investigators (good citizens all) who worked together to amass a growing wealth of factual knowledge without peddling allegiance to any particular chemical philosophy.[25] A similar depiction fits the rhetorical stance taken by Macquer's close collaborator and colleague, Antoine Baumé (1728– 1804) as well as by Torbern Bergman (1735–1784), Guyton de Morveau (1737–1816), and Antoine François Fourcroy (1755–1809) prior to their adoption of "revolutionary" chemistry, and by Joseph Priestley (1753– 1804), even well after the appearance of Lavoisier's assertively systematic *Traité élémentaire de chymie* in 1789. In Priestley's words,

> . . . our guessing at the system may be some guide to us in the discovery of the facts; but at present let us pay no attention to the system in any other view; and let us mutually communicate every new fact we discover, without troubling ourselves about the system to which it may be reduced . . . when all the facts belonging to any branch of science are collected the system will form itself.[26]

24. For a historical survey see A. M. Duncan, "Some Theoretical Aspects of Eighteenth-Century Tables of Affinity," *Annals of Science* 18 (1962): 177–194 (Part I); 217–232 (Part II).

25. Wilda Anderson, *Between the Library and the Laboratory* (Baltimore: Johns Hopkins University Press, 1984). Anderson also demonstrates Macquer's inability to remain neutral despite his rhetorical stance, which accounts for the omnipresent tension in his work.

26. Joseph Priestley, *The History and Present State of Electricity with Original Experiments* (London, 1767), 579–580. This ideal percolated to ranks of lesser authors, demonstrating its pervasive influence. For an example of such acceptance, see Sigaud de la Fond, *Description et usage d'un cabinet de physique expérimentale* (Paris, 1775), xvi–xvii.

Given Priestley's historical depiction as "revolutionary" chemistry's major antagonist and phlogiston's most ardent protagonist, such a pronouncement might at first seem ironic. But if we focus for a moment on the word "fact" and consider what was variously taken to constitute a fact by different actors, how facts were related to other facts and to nature as a whole, the inherent directedness of any apparently empirical endeavor becomes manifest.[27] Examining a series of synoptic tables and their declared significance from this period not only affords this general point's reiteration, it draws out the specific history of chemistry's discipline as well. We find that chemists generated and cataloged facts in synoptic tables throughout this period in the name of facilitating nature's own revelation. Their manner of proceeding gave rise to two simultaneous and ultimately irreconcilable outcomes. As organized sets of manipulative practices and their results, individual synoptic tables embodied explicitly directed systems of artifactual knowledge. As records of observed effects, these same tables cut a path across the world's surface, exposing relations among individual phenomena without determining the causal structure of the whole. In the context of chemists' rhetorical dedication to theoretical neutrality, there was no possibility of mediating between increasing practical prowess and the urge to encompass knowledgeably the world of nature. On the eve of chemistry's revolution, the chemical community found itself at an impasse, unable to unify theory and practice.

After a brief and fallow period,[28] interest in organizing distinctly chemical knowledge began climbing toward a new peak in the 1750s. Evidence for this can be seen generally in the growing popularity of public lecture courses and in the proliferation of journal articles, elementary textbooks, and chemical manuals. More specifically, we might look for a moment at the presentation of chemistry in the *Encyclopédie*, a work dedicated not only to general enlightenment but also to unifying the work of the hand and mind into a single system of knowledge.[29]

27. For a discussion of Priestley's views on the construction and definition of a fact, see a number of articles by John McEvoy, including his "Joseph Priestley, 'Aerial Philosopher': Metaphysics and Methodology in Priestley's Chemical Thought from 1762–1781," *Ambix* 25 (1978): 1–55, 93–113, 153–175; 26 (1979): 16–38.

28. Duncan (note 24) cites only one affinity table published between 1718 (Geoffroy's) and 1749.

29. This is indicated by the work's full title which describes it as a "reasoned dictionary of science, arts and crafts." For details, see Diderot's *Interprétation de la nature* and his article "Encyclopédie" as discussed in Lissa Roberts, "From Natural Theology to Naturalism: Diderot and the Perception of Rapports" (Ph.D. diss., University of California, Los Angeles, 1985). For a general discussion of chemistry in the *Encyclopédie* see Jean Claude Guedon, "The Still Life of a Tradition: Chemistry in the *Encyclopédie*" (Ph.D. diss., University of Wisconsin, 1974).

The primary author of chemistry articles for the *Encyclopédie* was G. F. Venel (1723–1775). Under the entry "chymie," Venel provided a programmatic statement of what he considered chemistry's ideal disciplinary structure.[30] In contrast to the mathematician d'Alembert, who defined chemistry as a technical art in the *Encyclopédie*'s "Discours préliminaire,"[31] Venel explicitly labeled chemistry as a science worthy of its own foundational principles. Declaring that it might actually take a revolution to establish chemistry along such autonomous lines, Venel sought meanwhile to demarcate chemistry's uniqueness of content and practice. He blamed the mathematical orientation of contemporary physics for having homogenized the material world to the point where all phenomena might be explained by preestablished mechanical principles and all bodies treated simply as qualitatively undifferentiated masses. Chemistry, if extricated from physics' suzerainty, posed the opportunity instead to explicate the world in terms of its sensibly obvious, heterogeneous fullness.

To this end chemistry's practitioners would be required to content themselves for the interim with a slow accumulation of experimentally garnered facts, as yet uninformed by mathematical laws of motion or other abstract principles. In place of philosophical or ideational cohesion, Venel suggested organization around the poles of experimental process and social practice. Every chemical reaction, he argued, reduced to a play between *rapports* and dissolution. The world of chemistry, then, encompassed two primary operations, the separation and union of material substances; and employed two general agents, heat and solvents. Here were the basic ingredients for a full-blown system of practical and ultimately theoretical chemistry. In order to be fleshed out, however, laboratory workers, accustomed to operating according to the intuitively bred wisdom of tacit craft knowledge, and more philosophically minded investigators, educated to follow trains of logic rather than material reactions, would have to join hands to shape their discipline as a reasoned outgrowth of laboratory activity.

To supplement the *Encyclopédie*'s textual presentation of chemistry, a

---

30. Diderot and d'Alembert, eds., *Encyclopédie* (Paris, 1753), 3: 408–437. In a longer monograph one might examine the individual differences between chemical commentators such as Venel and Macquer. Here, the important point—one which individual differences should not be allowed to overshadow—is their shared commitment to the notion of a cooperative community of chemical practitioners working toward the firm establishment of chemical knowledge.

For a discussion of their theoretical differences, see Rhoda Rappaport, "G. F. Rouelle: Eighteenth-Century Chemist and Teacher," *Chymia* 6 (1960): 68–99.

31. Jean LeRond d'Alembert, *Discours préliminaire* (Paris, 1929), 155.

table of *rapports* was reproduced in one of the accompanying volumes of plates, published in 1763. In terms of presentation, one might argue that it differed little from Geoffroy's 1718 table. Guyton de Morveau, for example, described the table in his *Elémens de chymie* as basically Geoffroy's with a few modifications made by the chemist Rouelle.[32] But such an interpretive view fails to take into account the immediate, graphic context of the table's presentation. The table, as a set of recorded results, is presented under the entry "chymie, laboratoire." It sits on the page directly beneath a detailed engraving of the very sort of chemistry laboratory in which the type of results it catalogs might be obtained. The table and engraving are clearly related, then, both by virtue of their spatial proximity and subject matter. This strategic juxtaposition of images served pictorially to ground organized chemical knowledge in the laboratory, depicting science as the slowly accruing outcome of cooperative labor.

We know that Diderot actively oversaw the layout and production of many of the *Encyclopédie*'s plates. We know too of his incisive aesthetic theories and art criticism.[33] Would it be too much to combine these two points and allow ourselves to be directed toward observing and analyzing the specific moment portrayed in this plate? Diderot valorized artists whose tableaux presented what might be called the "pregnant moment."[34] He claimed that by freezing time, as it were, just prior to a story's full denouement—at the moment when its depicted subjects are most totally absorbed in the action at hand—artists such as Jean-Baptiste Greuze and Jacques-Louis David invited the observer to catch their paintings' subjects "in the act." When such a pictorial move is successful, a painting's self-contained nature (its subjects display no awareness of a world outside the tableau's boundaries) and suspended action spur the observer to consider what is to follow; it catalyzes the viewer's own observational sense of active participation and naturalizes the depicted scene.

Returning to the picture at hand, we find that the depicted laboratory is

32. Guyton de Morveau et al., *Elémens de chymie* (Dijon, 1777), 1: 90.

33. For the standard biographical account, see Arthur Wilson, *Diderot* (New York: Oxford University Press, 1972). For a full account of Diderot's aesthetic views and writings, see Jacques Chouillet, *Formation des idées esthétiques de Diderot* (Paris: Armand Colin, 1973).

34. This is actually Lessing's phrase. For an analysis of Diderot's aesthetic views and art criticism along these interpretive lines, see Michael Fried, *Absorption and Theatricality: Painting and Beholder in the Age of Diderot* (Los Angeles: University of California Press, 1980). To place these views in a broader historical context, see idem, "Thomas Couture and the Theatricalization of Action in Nineteenth Century French Painting," *Artforum* 8 (1970): 36–46, and Svetlana Alpers, "Describe or Narrate? A Problem in Realistic Representation," *New Literary History* 8 (1976–1977): 15–41.

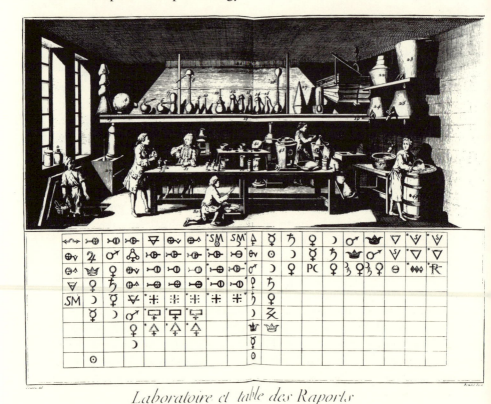

Laboratoire et table des Raports

Plate II. "Chymie, Laboratoire." Diderot, ed., *Encyclopédie: Recueil de planches, sur les sciences, les arts libéraux, et les arts méchaniques* vol. III (Paris, 1763).

populated not only by a collection of well labeled instruments but by an actively cooperative group of workers as well. Gentlemen and laborers alike (judging from their dress) jointly participate within its walls in the doing of chemistry. So absorbed are they in their work that they are oblivious to being observed and depicted. The outcome of their labor is pictorially undefined. But their activity is clearly directed toward some end; a certain orchestration of activity is evidenced by the obviously displayed division and coordination of labor, eye contact and communication between the workers. What is more, this "pregnant moment" of laboratory activity sits directly above a synoptic table which presents an equally well orchestrated set of relations derivable from the very sort of activity the picture presents.

The observer's eye moves between organized activity and organized results, the presentation of each buttressing the other in a linked portrayal of manipulative power and cooperation.[35] If this diptych has a moral (as was the case generally with the genre of tableaux interpreted by Diderot to which it is here compared), it was that the attainment of systematic scientific knowledge depended on the active cooperation of a community of investigators who, through an indissoluble marriage of hand and mind, directed their efforts toward fully uncovering a particular set (or sets) of experimental matters of fact.

In 1777, Guyton de Morveau, Hugues Maret (1726–1786), and Jean Durande (1733–1794) published their *Elémens de chymie théorique et pratique*, which included a reprint of the affinity table previously presented in the *Encyclopédie*. In this case, however, the table was put forward unaccompanied by a depiction of its laboratory context of discovery. This pictorial suppression parallels the book's expressed view of material composition. Guyton and his colleagues clearly distinguished between "chemical" elements, which they defined as primary only in the artifactual sense that chemists could not further decompose them, and "natural" elements, which they defined as truly elemental though manipulatively unisolatable.

According to these authors, the difference between chemical and natural elements had ultimately to do with the criteria by which each was theoretically individuated. Chemical elements were tangible, qualitatively distinguished by their sensible qualities and the observably distinct effects to which they gave rise. Natural elements, on the other hand, were considered homogeneous in composition and distinguished theoretically by quantitatively physical differences of figure, density, and porosity. As such, they constituted a more primary category of determination; observable relations among apparently heterogeneous substances had ultimately to be reduced to this underlying "reality." Affinity tables, in this context, were only a first step toward grasping the true nature of material interaction, just as the laboratory became only the context of their occasional facilitation.[36] For these authors, affinity chemistry ultimately gave way to the prowess of physics whereby, in keeping with their view that affinities were modifica-

35. My colleague Rob Hagendijk tells me that just this sort of juxtaposition of complementary and mutually reinforcing images is a standard and quite effective device in modern advertising and documentary reportage.

36. "L'homme n'a pas ces moyens à sa disposition, mais il a des procédés pour les mettre en jeu; c'est là tout ce qu'il peut, il ne fait par lui-même: ce n'est pas lui qui brule, qui dissout, qui crystallise; il met la nature dans la nécessité d'opérer ce qu'il desire." Guyton, 14.

tions of the universal force of gravity, elementary shapes might be mathematically extrapolated from the heterogeneous *rapports* observed in the laboratory.[37]

Guyton and his colleagues therefore organized their chemistry text instead around what they claimed as the fundamentally chemical process of dissolution. Although arguing that any organizational system "is nothing more than an imaginary series to aid the memory in assembling ideas and facts which are often disparate," they simultaneously argued that chemistry "never operates except by dissolution, whether to analyze a body or produce a new compound . . . dissolution is the thread which will lead us through this new quarry."[38] Accordingly, such an approach bore the Janus-like features of theoretical agnosticism (its organizational principle was directed explicitly toward human use) and of facilitating natural revelation by providing a practical thread for chemists to follow as they proceeded to weave experimentally through individual links in nature's chain of chemical being. It was hoped that by constructing and properly ordering a chain of chemical dissolutions in the laboratory, chemists might be led to expose the chain of nature as a whole.

It was with this end in mind that Guyton, Maret, and Durande included a synoptic table of solvent activity in their book. The table's grid was structured to accommodate every solvent chemists then knew and employed, without regard to substances' elemental nature. Given chemistry's proclaimed sensible status and the authors' acceptance of a more primary, though nonapprehensible, physical level of material determination, such an approach was perfectly reasonable; all chemical substances were regarded as truly compound, even the most seemingly primary of solvents, fire. Further, by constructing their table in this way, the authors directed it simultaneously toward two purposes. It provided, at a glance, the full accumulation of this genre of chemical facts uncovered thus far. Moreover, it offered "as upon the grid of a world map, the lands we have yet to discover."[39] Without predetermining the specific outcome of future operations, the table delimited and revealed particular areas as yet unexplored as well as the coordinative means and components of their exploration.

Since solvents embodied theory for these authors only insofar as theory

---

37. Ibid., 50–88. This conception of affinities was based on the ideas of Georges Buffon. See Buffon, *Natural History* (London, 1767) 10: 356–366 (originally published *Histoire naturelle des animaux*).

38. Guyton, 3.

39. Ibid., 89.

was equated with organized laboratory method, the outcome of each application had still to be empirically discovered. Guyton and his coauthors thus projected chemistry as a science limited to the sensible realm, concerned with and manipulative of observable qualities, operations, and products. However, they considered that nature cut a path through the laboratory, leaving a revelatory trail of truths. It was up to chemists to extract these "truths" from the manipulative situations in which they occurred, to translate and map them onto a spatial grid of chemical tabulation—a map of past discovery, future exploration, and manipulation.

Turning to contemporary considerations of affinity tables, we find that as early as 1763 Antoine Baumé suggested dividing them to record data obtained in the dry way (through application of heat) and in the wet way.[40] He argued that synoptic tables might thereby better reflect the growing intricacy and specificity of laboratory practice. Torbern Bergman was the first actually to organize a table along these lines. He went even further by offering one table for "single elective attractions" which recorded the relations within sets of only two substances, and another which recorded more complex operational relations.[41] We will examine each in turn and see what informed their structure and content.

Bergman's "table of single elective attractions" was a grand affair, recording many more relations than any affinity table before it. On one hand, this reflected Bergman's resignation to what he took as the unbridgeable gap between laboratory activity and natural knowledge. On the other hand, as we will see, it pointed to the privileged status he accorded laboratory phenomena over the very activity of nature in the attainment of chemical knowledge about the world.

Bergman cautioned his readers against assuming "that the methods we observe and employ are the methods employed by nature itself." Claiming ignorance regarding matter's true elemental composition and the ultimate cause of elective attraction, he offered distinctively practical criteria for inclusion in his table.[42] Bergman distinguished substances in terms of their observably apparent, unique, and constant properties and their penchant to combine with certain other substances. Such criteria completely overshad-

40. Antoine Baumé, *Manuel de chymie* (Paris, 1763). See also idem, *Chymie expérimentale et raisonnée* (Paris, 1773).

41. See Torbern Bergman, *A Dissertation on Elective Attractions* (London, 1785). For his discussion of wet and dry analysis, see ibid., 16–17.

42. Bergman, *Physical and Chemical Essays,* trans. Edmund Cullen (London, 1788), xxix, and *An Essay on the Usefulness of Chemistry and Its Application to the Various Occasions of Life,* trans. Jeremy Bentham (London, 1783), 1–2.

owed the question of a body's elemental nature, identifiably situating substances in the relational terms of his table's grid of presentation. Indeed, Bergman wrote:

> Should they [chemical substances] be derived from others, they ought not, on this account, to be excluded, for they are now different, have constant properties, exercise their attractive powers without decomposition, and can at pleasure be obtained perfectly alike.[43]

Bergman considered his manner of tabulating facts of elective attraction as the "key to the whole science" of chemistry.[44] It allowed him to record relations reflective of his view "that the same elements, combined in different manners, or in different proportions, may produce different bodies."[45]

The table's structure of presentation, then, reflected Bergman's trust in laboratory circumstances as more knowledgeably fruitful than nature. Though unable to (re)create nature's original materials, chemists could construct replicable manipulative spaces at will, within which bodies might be combined in apparent imitation of nature or with results entirely unknown in the natural world. Thus assured of their own productive capabilities, even if bodies' "true" composition and nature might not be uncoverable, chemists might nonetheless gain exhaustive knowledge of chemical relations as conceived under laboratory conditions.[46]

The complexity this entailed is portrayed in the symbolic language of Bergman's "table of double elective attractions." In the tradition of Geoffroy, Bergman elected to employ a system of chemical symbols for his table rather than to rely on words,[47] but he geared his symbolic language to do

43. Bergman, *Elective Attractions*, 71–72.

44. Bergman, *Essay*, 15.

45. Ibid., 139. For example, Bergman argued that experiments had demonstrated that, depending on the particular (quantitative) combination of pure air and phlogiston, depraved air, heat or light was produced.

The anonymous London translator superimposed on Cullen's translation of Bergman's *Physical and Chemical Essays* wrote a Bermanesque footnote on how to conceive of substances that combine in different proportions to form different compounds: "a compound becomes in this respect an element, and unites with one of its constituent parts, in order to form a new compound" (14–15).

46. Bergman, *Essay*, 75–76, 137–139. See also *Physical and Chemical*, xxxii.: "With respect to facts, indeed, which are collected by experiment, no dispute can arise, as they may at pleasure be appealed to, and considered in every point of view with sufficient care. The case is otherwise with those circumstances which depend solely upon the operations of nature; for these, if the fit time be neglected, do not again occur, but accidentally, or perhaps after an interval of years."

47. Contemporaneously, Guyton de Morveau argued against using symbols to represent chemical substances. Not only did they symbolically tie chemistry to the very occult, alchemical past from which current reformers were trying to extricate it, he claimed, but from a purely practical perspective, they were both difficult to explain and easy to forget. See his *Elémens de chymie*, 91.

more than depict single elective attractions between individual substances. Bergman's system enabled him to portray the *processes* entailed in complex reactions involving double elective attractions. In his hands, symbolic art— which opened the world to multitudinous layers of feeling and interpretation—was totally co-opted by the single-minded purpose of science. Bergman managed to subdue the hieroglyphic nature of chemical symbols,[48] fashioning them instead as instruments which were no more alchemically or poetically evocative than any other instrument at the chemist's disposal. Their sole job was to represent a given chemical reaction from start to finish.

The following explication of one such symbolic scheme exemplifies the exhaustive nature of Bergman's symbolic language.

SCHEME 20. pl. 1 exhibits the decomposition of calcareous hepar by the vitriolic acid. On the left side appears the hepar, indicated by the signs of its proximate principles united; but within the vertical bracket these principles are seen separate, one above the other. On the right, opposite the calcareous earth, is placed the sign of vitriolic acid; in the middle stands the sign of water, intimating that the three surrounding bodies freely exercise their attractive powers in it. Now, as vitriolic acid attracts calcareous earth more forcibly than sulphur does, it destroys the composition of the hepar; the extruded sulphur being by itself insoluble falls to the bottom, which is signified by the point of the lower horizontal half-bracket being turned downwards; and as the new compound, vitriolated calcareous earth [gypsum], also subsides, unless the quantity of water be very large, the point of the upper bracket is likewise turned downwards. The complete horizontal bracket indicates a new combination, but the half-bracket serves merely to shew [sic] by its point whether the substance from which it is drawn remains in the liquor, or falls to the bottom. The absence of horizontal brackets indicates that the original compound remains entire. . . . Those operations which are performed in the dry way, are distinguished by the character of fire, which is placed in the middle.[49]

The force of Bergman's symbolic representation was that it captured the whole of a given operation within the confines of its symbolic borders, mirroring the instrumental confines of the laboratory vessels within which

48. Little has been written on the popular interest in hieroglyphics and their relation to other languages of artistic expression in the eighteenth century. Less still (if anything) has been written on the link between this preoccupation and the revivification of symbolic systems in chemistry. See Madeleine V.-David, *Le aébat sur les écritures et l'hiéroglyphe aux XVVIIe et XVIIIe siècles* (Paris: Sevpen, 1965); Jacques Derrida, "Scribble (pouvoir/écrire)", preface to William Warburton, *L'Essai sur les hiéroglyphes* (Paris: Aubier-Flammarion, 1978), originally published as *The Divine Legation of Moses Demonstrated* (London, 1742); and Lissa Roberts, "From Natural Theology to Naturalism," chap. 3.

49. Bergman, *Elective Attractions*, 12–13.

+ ⊕ Vitriolic acid

☿p Pure calcareous lime

▽ Water

♄ Sulphur

Plate III. "Table of Double Elective Attractions: Scheme 20." Torbern Bergman, *Dissertation on Elective Attractions* (London, 1785). Reprinted by permission of the publishers from *A Source Book in Chemistry 1400–1900,* edited by Henry M. Leicester and Herbert S. Klickstein. Cambridge, Mass.: Harvard University Press, 1952. Copyright © 1952 by the President and Fellows of Harvard College.

the complex reaction took place and making the entire process observable at a single glance. In contrast, discursive language alienated the reader from actual occurrences by spreading out across the page to describe activity one discrete step at a time.[50] Just as the "natural" language of pictorial representation was seen as enabling a painter to represent artistically the simultaneity of worldly occurrence by appealing to the observer's intuitive apprehension of a given ensemble, Bergman offered his system as a means of naturalizing the representation of chemical reactions. Through increasing familiarity with his language's representative signs, chemists might acquire an intuitive grasp and appreciation of the complex nature of chemical phenomena.[51]

Yet, while disciplining the presentation of laboratory chemistry's observational matters of fact and providing a directed outlet for chemists' ongoing investigations, how far did Bergman's table—or any other table from this period—go toward projecting chemistry as a systematic science capable of knowledgeably encompassing the world? By the late 1770s to the early 1780s an increasing number of chemists began to articulate discontent with the extent to which their field's knowledgeable aspirations were being fulfilled. Their voiced concerns, however, took them in a direction that

50. This is an interesting contrast to Robert Boyle, who discursively filled his experimental reports with as much detail—both ambient and directly experimental—as possible. Boyle claimed that such an exhaustive presentation naturalized his report to the point that readers need not actually repeat the original experiment to judge of its truth. See Robert Boyle, "New Experiments Physico-Mechanical, Touching the Spring of the Air" (London, 1660), 1–2. See also Shapin and Schaffer, *Leviathan,* 60–65, especially 62, note 84.

51. My reliance on contemporary theories of aesthetic apprehension is intended to help set such tabulating endeavors in the context of their own time.

remained consonant with the general direction their discipline was taking. Rather than focus directly on the question of chemical knowledge *per se,* their commentaries and actions took two other primary directions, focusing either on criteria for membership in the chemical community (properly organized behavior) or on the need for language reform (properly organized expression).

The social issue of membership criteria is certainly crucial to the history of chemistry's disciplinary development, but its story's full telling is unfortunately beyond the limited scope of this essay. Some brief comments are nonetheless appropriate. A survey of the literature from and about this period indicates that the rites and rights of membership related to an interplay between three factors: perceived manipulative abilities and technical acumen in laboratory settings; acceptance and use of polite, theoretically neutral discourse (that is, in the sense of not asserting an overall system) for communication in general and experimental reporting in particular, and success in situating oneself in a recognized network of active participants.

As Carleton Perrin has shown, it was the last of these factors that loomed largest in the critical period of chemistry's revolutionary debates in the late 1780s and 1790s.[52] Prior to that, a marriage between the first two received more vocalized attention. Thus, on one hand, Priestley might comment on what he considered the undue complexity, costliness, and awkwardness of some of Lavoisier's experiments and yet welcome him as a cooperative, gentlemanly colleague.[53] On the other hand, he advocated that the chemist and entrepreneur Bryan Higgins be banned from the recognized cadre of chemists. Not only was Higgins technically maladroit, Priestley argued. He was a sleazy social operator as well; the man had actually stooped to charging Priestley (quite maliciously, Priestley felt) with plagiarism![54]

More central to the presentation at hand is the question of language

52. Carleton Perrin, "The Triumph of the Antiphlogistians," in Harry Woolf, ed., *The Analytic Spirit: Essays in the History of Science in Honor of Henry Guerlac* (Ithaca, N.Y.: Cornell University Press, 1981) 40–63. Controversy during this period was increasingly adjudicated by the blatant use of strategies such as championing the new system of nomenclature, restructuring the Académie des Sciences, capturing or creating journals as organs of publicity and persuasion, and so on.

53. Joseph Priestley, *Experiments and Observations on Different Kinds of Air* (London, 1775) 1: 193–194.
Negative myths to the contrary, Lavoisier was highly respectful of Priestley the chemist as well. For a good deal of evidence see F. L. Holmes, *Lavoisier and the Chemistry of Life* (Madison: University of Wisconsin Press, 1985), passim.

54. Joseph Priestley, "Philosophical Empiricism," appended to *Experiments and Observations*, vol. I.

reform. Bergman, among others, worried that potential advancement in chemical knowledge was being sacrificed to the continued veneration of chemistry's traditional, often obscure and confused, language. He asked, "Why, then should chemistry, which examines the real nature of things, still adopt names, which suggest false ideas, and savour strongly of ignorance and imposition?"[55] From 1775 until his death in 1784, Bergman suggested reforms in chemical nomenclature aimed specifically at the clarification of individual, experimental facts. In general, his idea was to label component substances (primarily acids and alkalis) with easily and quickly identifiable terms and then to name their compounds explicitly as their compounds. For example, Bergman wrote that "vitriolic acid" (previously termed "oil of vitriol") plus "vegetable fixed alkali" gave rise to "vitriolated tartar" or "more properly, vitriolated vegetable alkali."[56] Compound names thus embodied (potentially long) descriptive histories of their signified compounds' experimentally observed parentage.

In 1782, Guyton de Morveau published his "Mémoire sur les dénominations chymiques, la nécessité d'en perfectionner le système et les règles pour y parvenir."[57] Included in the memoir was an apparently new sort of chemistry table—a table of nomenclature—aimed at bringing order to chemistry's language just as contemporary synoptic tables were bringing order to its matters of fact. When the extent of these tables' similarity is fully exposed, however, it is seen that Guyton's table was novel only in a very limited sense.[58] True, it was a table of nomenclature and not one that directly charted material relations, but its structure was essentially the same.

Rather than presenting an exhaustive list of terms consequential to an initial set of organizing principles or core elements, the table provided an analogical base from which chemists might draw, in a case by case manner, to name other substances. The table begins with a list of acids divided according to what realm of nature (mineral, vegetable, or animal) they are observed to inhabit, followed by a corresponding column listing the categories of generic salts formed out of these acids' combination with any given base. These two columns are separated by a double line from a list of

55. "On the Investigation of Truth," prefatory essay preceding Bergman, *Physical and Chemical*, xiv.

56. Bergman, *Essay*, 88.

57. *Observations sur la Physique* 19: 370–382.

58. This view goes a long way toward explaining why Guyton's memoir met with such general approbation in contrast to the *Méthode de nomenclature chymique* (Paris, 1787) which he and his collaborators Lavoisier, Fourcroy, and Berthollet presented to the Académie des Sciences in 1787. The 1782 memoir offered new information without challenging the recognized structure of discourse, whereas the 1787 memoir, as will be argued shortly, implied nothing less than a wholesale revision of chemistry's disciplinary structure.

RÈGNE	ACIDE	Les Sels formés de ces Acides prennent les noms génériques de	BASES ou substances qui s'unissent aux Acides	EXEMPLES pour la classe des Variable	EXEMPLES pris de diverse classes
Des trois Règnes	Méphitique ou Air fixe	Méphites	Phlogistique	Soufre vitriolique ou soufre commun	Soufre méphitique ou Plombagine
	Vitriolique	Vitriols	Alumine ou Terre de l'argille	Vitriol alumineux ou Alun	Nitre alumineux
	Nitreux	Nitres	Calca ou Terre calcaire	Vitriol calcaire ou Sélénite	Muriate calcaire
	Muriatique ou du sel marin	Muriates	Magnésie	Vitriol magnésien ou Sel d'epsom	Acète de magnésie
Minéral	Régulin	Régulins	Barote ou Terre du Spath pesant	Vitriol barotique ou Spath pesant	Tartre barotique
	Arsenical	Arsenics	Potasse ou Alkali fixe végétal	Vitriol de potasse ou Tartre vitriolé	Arsénical de potasse
	Boracin ou sel sédatif	Borax	Soude ou Alkali fixe minéral	Vitriol de Soude ou Sel de Glauber	Borax de Soude ou Borax commun
	Fluorique ou du spath fluor.	Fluors	Ammoniac ou Alkali volatil	Vitriol ammoniacal	Fluor ammoniacal
			Or	Vitriol d'or	Régule d'or
			Argent	Vitriol d'argent	Oxalin d'argent
			Platine	Vitriol de platine	Saccharin de platine
			Mercure	Vitriol de mercure	Chlorat de mercure
			Cuivre	Vitriol de cuivre ou Vitriol de Chypre	Lignite de cuivre
			Plomb	Vitriol de plomb	Phosphate de plomb
			Étain	Vitriol d'étain	Formicin d'étain
			Fer	Vitriol de fer ou Couperose verte	Sébate martial
Végétal	Acéteux ou Vinaigre	Acètes	Antimoine (ou ... de Régule d)	Vitriol antimonial	Muriate antimonial ou Beurre d'antimoine
	Tartareux ou de la Tartre	Tartres	Bismuth	Vitriol de bismuth	Galacte de bismuth
	Oxalin ou de l'Oseille	Oxalins	Zinc	Vitriol de zinc ou Couperose blanche	Boux de zinc
	Saccharin ou du Sucre	Saccharins	Arsenic	Vitriol d'arsenic	Muriate d'arsenic
	Citronin ou du Citron	Citrates	Cobalt	Vitriol de cobalt	Saccharin de cobalt
	Ligneux ou du Bois	Lignites	Nickel	Vitriol de Nickel	Formicin de Nickel
			Manganèse	Vitriol de manganèse	Oxalin de manganèse
Animal	Phosphorique	Phosphates	Esprit-de-vin	Esther vitriolique	Esther liquique ou Ether de Goëtling, &c. &c. &c.
	Formicin ou des Fourmis	Formicins			
	Sébacé ou du Suif	Sébates			
	Galactique ou du Lait	Galactes			

Les soufres & les libers deviennent aux-mêmes noms de genres, & se distinguent par l'espèce de l'acide.

Les noms de ces bases, en leur adjectif, ajoutés aux substantifs qui indiquent les genres des acides, forment les dénominations exactes, comme on le voit dans les exemples suivans.

Les dix-huit acides, les vingt-quatre bases & les produits de leur union, forment cinq quatre cent soixante-quatorze dénominations claires & méthodiques, indépendamment des vapeurs ou composés à trois parties, dont les noms viennent encore dans ce système, comme vapeur de foie, hépar ammoniacal, esprit d'arpus, &c. &c.

N. B. Lorsque les acides particuliers déjà entrevus dans la molybdène, l'étain, &c. (comme plus connus), on en formera les noms d'acide molybdique & molybdés, d'acide stannique & stannés, &c. Il en sera de même des nouvelles bases. Le nouveau demi-métal trouvé par M. Bergmann dans les fers cassans, pourra être nommé sydéroïde, caché dans le fer.

Plate IV. "Tableau de nomenclature chymique, contenant les principales dénominations analogiques, et des examples de formation des noms composés." Guyton de Morveau, "Sur les dénominations chymiques," *Observations sur la physique, sur l'histoire naturelle et sur les arts et métiers* 19 (1782): 370–382. Reprinted by permission of the publishers from Wilda Anderson, *Between the Library and the Laboratory.* Baltimore: Johns Hopkins University Press, 1984.

bases known to unite with acids and two further columns: one that lists the compound names resulting from the union of vitriolic acid with each of the listed bases, and one other that offers examples of compound names drawn from the union of a variety of acids with each successively named base. In all cases, compound names are just that: names composed out of the names of a compound's empirically observed constituents.

Just as Bergman proposed his tables as founded on a utilitarian criterion, so did Guyton justify nomenclatural reform as a needed response to the contemporary proliferation of newly revealed and/or accepted substances. It was simply becoming too difficult to bear in mind all that the un-systematically growing field of chemistry entailed. And, just as Bergman had transformed chemical symbols, previously redolent with alchemical significance, into brute instruments of empirical recording and manipulation, so did Guyton describe his terms as conventional markers. Indeed, he specifically advocated that chemists rely on names with as little meaning in ordinary language as possible.[59] Claiming that his terms expressed only "the abstract objects of my thought without demanding definitions,"[60] Guyton further asserted that any simple chemical term was distanced from the substance it signified by the simple historical fact that a name was always given to a substance long before the substance's character was exhaustively explored and known. Finally, just as Bergman's tables cataloged an empirical accrual of observed affinities and directed chemists toward future research without predetermining the specific outcome of individual cases, Guyton's table of nomenclature (which, like Bergman's work, focused primarily on acid-alkali relations) offered a formula for naming compounds explicitly as the products of empirically observed combinations. Investigators had still to rely on empirical findings as they followed the nomenclatural direction he provided. Like Bergman's, Guyton's tabular and nomenclatural ideal was to embody the narrative history of experimentally revealed relations.

The Table Is Set

When Guyton de Morveau and his colleagues Antoine Laurent Lavoisier (1743–1794), Antoine François Fourcroy (1755–1809), and Claude Louis

59. ". . . we should at least prefer a name which is further removed from common use, because it is much better if the technical terms of a science express nothing which is known and recall no idea than indicate false resemblances which lead beginners astray and always astonish the most educated people." Guyton, *Memoir* (1782), quoted in Maurice Crosland, *Historical Studies*, 154. See also W. Anderson, *Library*, 175, note 165.

60. Adoption of his nomenclature thus entailed no deeper theoretical commitments. See *Observations sur la Physique* 18 (1781): 69ff. Quoted in Crosland, *Historical Studies*, 156.

Berthollet (1748–1822) published their *Méthode de nomenclature chymique* in 1787, they operated with a very different view of language and its relation to knowledge formation. Until this point, language reform in chemistry had been piecemeal (Guyton's 1782 memoir notwithstanding), directed toward the clarification of individual "matters of fact." The ideal was to generate a language (much as chemists generated experimental data) to which all chemists could accede, whatever their private views. But now, as Lavoisier noted, this seemed an unrealizable goal.

> . . . in a science which, in a certain sense, is on the move, which is taking great strides toward its perfection, in which new theories are presenting themselves, it was extremely difficult to form a language amenable to different systems and which satisfied everyone's opinions without adopting a single one.[61]

With this realization, the challenge, as Lavoisier described it, was to embody linguistically the one system that provided "a faithful mirror of what nature presents us."[62] Buttressed by his reading of Condillac, Lavoisier argued further that chemistry's language of choice had to be one whose terms for what was known contained the seeds of what was yet unknown. It must be a language—"a method of naming rather than a nomenclature"—that "marks in advance the place and name of new substances which might be uncovered."[63] But how might a language and the chemical system it articulated simultaneously provide a passive mirror image of nature and serve as an actively structuring method of discovery? As the following analysis of the *Méthode*'s table of nomenclature and two representative tables from Lavoisier's *Traité élémentaire* is intended to show, this apparent contradiction was resolved by identifying nature from the outset with both the process and product of (humanly engineered) chemical manipulation.

The full implications of collapsing natural knowledge into manipulative practice were not realized until the publication of Lavoisier's *Traité*. The *Méthode*'s table of nomenclature nonetheless shows the direction "revolutionary" chemistry was taking. Unlike previous tables examined here, this table begins with a column naming "*substances non decomposées*," its content informed by explicit criteria of what constituted a simple substance.[64] The five subcategories of simples can be broken down into two. The first group

61. *Méthode*, 4.
62. Ibid., 14.
63. Ibid., 16, 17.
64. The table is reproduced in Anderson, *Library*, 154–155.

was described in the text as composed of substances that appeared most definitely to resist analysis and that simultaneously were most active in combining with other substances.[65] It was given no identifying label in the table, accentuating the implication that its categorical structure and existence were natural and not humanly construed. Although thus considered the most fundamental elements of nature, none of its constituents (light, caloric, oxygen, and hydrogen) had ever been isolated; they were "known" only in compound form or inferentially from the effects that chemists traced back to what was theoretically (and nominatively) asserted as their causal presence. The other four categories (acidic bases, metals, earths, and alkalis) were ultimately delineated in terms of the first category's action upon them.[66] They were deemed simple because, as yet, they had resisted decomposition at the hands of chemical investigators.

All in all, the determining criteria for inclusion in the table's first column and for the type of data charted in the table's other columns had to do with activity—the directed activity of chemists in their laboratories and/or that ascribed to elements considered so primary and pervasive in their action to escape isolated detection. What must be noted is that the direction thus given chemistry (and excluded from it) and depicted in this table resulted from a combination of chemists' instrumental abilities and their theoretical construction of the key elements oxygen and caloric. Oxygen and caloric were both defined and named exclusively in terms of a set of phenomena to which they purportedly gave rise. Any reference to them whatsoever threw chemical investigators immediately back out into the world of effects. But it was a world now structured by what was claimed as oxygen and caloric's pervasively causal action on a set of substances construed as simple in the context of the laboratory. Readers of the *Méthode* were offered only the choice of rejecting its system wholesale "or irresistibly following the route it has marked out."[67]

It was in the tables of Lavoisier's *Traité élémentaire* that the all-encompassing nature of oxygen and caloric's proclaimed power—and, ultimately,

65. *Méthode*, 28.
66. Thus acidic bases, not all of which had been isolated (for example, the muriatic radical), combined with oxygen to form acids, whereas metals combined with oxygen first (and sometimes only) to produce oxides. With one exception (ammonia gas), earths and alkalis were conspicuous for the absence of any recorded data about them in the table—a foreshadowing of their fate in the *Traité élémentaire*.
67. *Méthode*, 12. This point was not lost on dissenting chemists such as Joseph Black who complained that "to accept the new nomenclature was to accept the new French theories." Quoted from Henry Guerlac, *Antoine Laurent Lavoisier: Chemist and Revolutionary* (New York: Charles Scribner's Sons, 1975), 108.

the power of chemists' own manipulative abilities—was fully and finally manifested; oxygen by its blatant omnipresence, caloric by its thorough instrumentalization, human manipulation by the way it defined, encased, and directed these and all other chemical elements. In the *Traité* a related series of tables replaced the *Méthode*'s table of nomenclature in which a fully organized world of effects existed more implicitly than explicitly. Beginning with a table of elements and simple substances, Lavoisier's work branches out into a series of other tables, the order and content of each determined by the causal power purportedly embodied in one of the first table's primary elements. After a discussion of Lavoisier's table of elements and simple substances, his "table of the binary combinations of oxygen with simple substances" will be presented as the prime example of "revolutionary" chemistry's organizing reach and power.

Like the 1787 table, Lavoisier's table of simple substances was divided into a small group of unisolable substances construed as so omnipresent and causally active as to "be considered as the elements of bodies"[68] and a larger class considered simple only because chemists had not yet succeeded in further decomposing them. This second class was now divided into three subcategories instead of four: oxydable and acidifiable simple nonmetallic substances, oxydable and acidifiable simple metals, and salifiable simple earthy substances. As can be gathered from the first two subdivisions' titles, their identities were marked from the outset by their combinative relation to oxygen. As to the salifiable simple earths, Lavoisier challenged chemists to investigate their nature, writing:

> They are the only bodies of the salifiable class which have no tendency to unite with oxygen; and I am much inclined to believe that this proceeds from their being already saturated with that element.[69]

Whether his hypothesis proved correct or not, this category too was definitionally prescribed by the causally active presence of oxygen.

Fixed alkalis, which had composed a fifth category in 1787, disappeared from Lavoisier's table because of widespread suspicion that they were compounds containing the substance azote. Symmetrically, azote was now elevated into the first class of elements. But even fixed alkalis were marked by oxygen's presence, or, in this case, lack of presence. Rather than rename the element azote "alkaligene" as suggested by Fourcroy, thus overtly

68. *Elements of Chemistry*, 175. Compare my discussion with that of Anderson, 138–146.
69. *Elements of Chemistry*, 177.

TABLE OF SIMPLE SUBSTANCES.

Simple fubſtances belonging to all the kingdoms of nature, which may be conſidered as the elements of bodies.

New Names.	Correſpondent old Names.
Light	Light.
Caloric	Heat. Principle or element of heat. Fire. Igneous fluid. Matter of fire and of heat.
Oxygen	Dephlogiſticated air. Empyreal air. Vital air, or Baſe of vital air.
Azote	Phlogiſticated air or gas. Mephitis, or its baſe.
Hydrogen	Inflammable air or gas, or the baſe of inflammable air.

Oxydable and Acidifiable ſimple Subſtances not Metallic.

New Names.	Correſpondent old names.
Sulphur	
Phoſphorus	The ſame names.
Charcoal	
Muriatic radical	
Fluoric radical	Still unknown.
Boracic radical	

Oxydable and Acidifiable ſimple Metallic Bodies.

New Names.		Correſpondent Old Names.
Antimony		Antimony.
Arſenic		Arſenic.
Biſmuth		Biſmuth.
Cobalt		Cobalt.
Copper		Copper.
Gold		Gold.
Iron		Iron.
Lead		Lead.
Manganeſe	Regulus of	Manganeſe.
Mercury		Mercury.
Molybdena		Molybdena.
Nickel		Nickel.
Platina		Platina.
Silver		Silver.
Tin		Tin.
Tungſtein		Tungſtein.
Zinc		Zinc.

Salifiable

Salifiable ſimple Earthy Subſtances.

New Names.	Correſpondent old Names.
Lime	Chalk, calcareous earth. Quicklime.
Magneſia	Magneſia, baſe of Epſom ſalt. Calcined or cauſtic magneſia.
Barytes	Barytes, or heavy earth.
Argill	Clay, earth of alum.
Silex	Siliceous or vitrifiable earth.

Plate V. "Table of Simple Substances." Antoine Laurent Lavoisier, *Elements of Chemistry*, translated by Robert Kerr (Edinburgh, 1790).

labeling it as the generative principle of alkalinity, Lavoisier coined the term "azuret" for any compound of azote and another simple substance. In his words, "to these we give the name azurets, preserving the termination in *uret* for all nonoxygenated compounds. It is extremely probable that all the alkaline substances may hereafter be found to belong to this genus of azurets."[70]

Lavoisier's table of simple substances, then, is thoroughly marked by the active presence of oxygen. The organized consequences of this are fully displayed in one of the *Traité*'s other tables, the previously mentioned "table of binary combinations of oxygen with simple substances." Unlike all previous tables discussed here, it is quite saliently organized by the power of a single theoretical assertion: oxygen, as its name declares, is the generative principle of acidity. Once placed on the table, all other substances take on a teleological identity consequential to this principle. A full world of chemical possibility spreads out across the page. Metallic simples, for example, combine with oxygen first to form oxyds. Rather than standing as calces which point back to their vulcanic origins, they now bear the mark of the "first degree of oxygenation." The observer's eye is thus directed onward as the organizing power of oxygen is manifested in the second, third, and fourth degrees.

As a surface made continuous by oxygen's omnipresence, the table allows no gaps or inconsistencies. That which is not yet known nonetheless has its position on the table reserved by a prefabricated name. This is true not only for compounds, but for simples as well. If an acid is determined *a priori* to contain oxygen, so too must it contain an acidic base. As the table shows, four such—as yet unisolated—radicals are listed in the column of simple substances, testifying to the ubiquity of oxygen's determinative strength.[71]

And now, what of caloric, so far conspicuous primarily for its absence? After its having been given such a prominent position in the 1787 table of nomenclature, it might seem strange that not a single table exists in the *Traité* to chart caloric's organizational impact on the world of chemistry. This omission is intimately related to another, equally striking, tabular omission. Although the *Traité* abounds with tables (thirty-three, not including the eight conversion tables in the Appendix), it does not contain a

70. Ibid., 195.
71. As François Dagognet has commented, "Lavoisier was not content, as were his predecessors, to place things in an order [a catalogue], but he posed or supposed that an order preceded the creation of things [a classification]" *Tableaux et langages de la chimie* (Paris: Editions du Seuil, 1969), 27 (my trans.).

TABLE of the binary Combinations of Oxygen with simple Substances.

Names of the simple substances.	First degree of oxygenation. New Names.	First degree of oxygenation. Ancient Names.	Second degree of oxygenation. New Names.	Second degree of oxygenation. Ancient Names.	Third degree of oxygenation. New Names.	Third degree of oxygenation. Ancient Names.	Fourth degree of oxygenation. New Names.	Fourth degree of oxygenation. Ancient Names.
Combinations of oxygen with simple non-metallic substances.								
Caloric	Oxygen gas	Vital or dephlogisticated air • •						
Hydrogen	Water •							
Azote	Nitrous oxyd, or base of nitrous gas	Nitrous gas or air	Nitrous acid	Smoaking nitrous acid	Nitric acid	Pale, or not smoaking nitrous acid	Oxygenated nitric acid	Unknown
Charcoal	Oxyd of charcoal, or carbonic oxyd	Unknown	Carbonous acid	Unknown	Carbonic acid	Fixed air	Oxygenated carbonic acid	Unknown
Sulphur	Oxyd of sulphur	Soft sulphur	Sulphurous acid	Unknown	Sulphuric acid	Vitriolic acid	Oxygenated sulphuric acid	Unknown
Phosphorus	Oxyd of phosphorus	Residuum from the combustion of phosphorus	Phosphorous acid	Volatile acid of phosphorus	Phosphoric acid	Phosphoric acid	Oxygenated phosphoric acid	Unknown
Muriatic radical	Muriatic oxyd	Unknown	Muriatic acid	Unknown	Muriatic acid	Marine acid	Oxygenated muriatic acid	Dephlogisticated marine acid
Fluoric radical	Fluoric oxyd	Unknown	Fluorous acid	Unknown	Fluoric acid	Unknown till lately		
Boracic radical	Boracic oxyd	Unknown	Boracic acid	Unknown	Boracic acid	Homberg's sedative salt •		
Combinations of oxygen with the simple metallic substances.								
Antimony	Grey oxyd of antimony	Grey calx of antimony	White oxyd of antimony	White calx of antimony, diaphoretic antimony	Antimonic acid			
Silver	Oxyd of silver	Calx of silver			Argentic acid			
Arsenic	Grey oxyd of arsenic	Grey calx of arsenic	White oxyd of arsenic	White calx of arsenic	Arseniac acid	Acid of arsenic	Oxygenated arsenic acid	Unknown
Bismuth	Grey oxyd of bismuth	Grey calx of bismuth	White oxyd of bismuth	White calx of bismuth	Bismuthic acid			
Cobalt	Grey oxyd of cobalt	Grey calx of cobalt			Cobaltic acid			
Copper	Brown oxyd of copper	Brown calx of copper	Blue and green oxyds of copper	Blue and green calxes of copper	Cupric acid			
Tin	Grey oxyd of tin	Grey calx of tin	White oxyd of tin	White calx of tin, or putty of tin	Stannic acid			
Iron	Black oxyd of iron	Martial ethiops	Yellow and red oxyds of iron	Ochre and rust of iron	Ferric acid			
Manganese	Black oxyd of manganese	Black calx of manganese	White oxyd of manganese	White calx of manganese	Manganic acid			
Mercury	Black oxyd of mercury	Ethiops mineral †	Yellow and red oxyds of mercury	Turbith mineral, red precipitate, calcined mercury, precipitate per se	Mercuric acid			
Molybdena	Oxyd of molybdena	Calx of molybdena			Molybdic acid	Acid of molybdena	Oxygenated molybdic acid	Unknown
Nickel	Oxyd of nickel	Calx of nickel			Nickelic acid			
Gold	Yellow oxyd of gold	Yellow calx of gold	Red oxyd of gold	Red calx of gold, purple precipitate of cassius	Auric acid			
Platina	Yellow oxyd of platina	Yellow calx of platina			Platinic acid			
Lead	Grey oxyd of lead	Grey calx of lead	Yellow and red oxyds of lead	Massicot and minium	Plumbic acid			
Tungstein	Oxyd of Tungstein	Calx of Tungstein			Tungstic acid	Acid of Tungstein	Oxygenated Tungstic acid	Unknown
Zinc	Grey oxyd of zinc	Grey calx of zinc	White oxyd of zinc	White calx of zinc, pompholix	Zincic acid			

* Only one degree of oxygenation of hydrogen is hitherto known —A

† Ethiops mineral is the sulphuret of mercury; this though I have been called black precipitate of mercury. —E.

Plate VI. "Table of the Binary Combinations of Oxygen with Simple Substances." Antoine Laurent Lavoisier, *Elements of Chemistry,* translated by Robert Kerr (Edinburgh, 1790).

single affinity table. As Lavoisier made explicitly clear in his preface[72] and through his reliance on their explanatory utility in his expositions of experiments, he never ceased to accept the central role of affinities in chemistry. But the structure of his chemical system placed them, along with caloric, in a very different context—one that accentuates how the world of Lavoisier's chemistry was ultimately structured by the determining force of human manipulation.

Included in the *Méthode* was a memoir by Jean Henri Hassenfratz and Pierre August Adet which presented a new system of chemical symbols to complement chemistry's new nomenclature.[73] In recognition of their belief in caloric's omnipresence in all other bodies, the authors presented a "table of simple substances combined with caloric, producing the solid, liquid and aeriform states." To avoid undue repetition, they claimed, Hassenfratz and Adet symbolically wrote caloric out of all substances in the state of solidity. The symbol for a substance in the state of solidity was the same as for the substance itself, while symbols for both the liquid and aeriform states included the overtly marked presence of caloric. Thus, in the name of "convenience" (this term is the authors'), caloric's move toward compositional transparence was institutionalized.

Indeed, given its theoretical construction, the only time caloric's dynamic presence could be noted was when a bodily change (either change of state or decomposition) was occurring. Lavoisier encased this activity by naming the machine he and Pierre Simon Laplace (1749–1827) had previously constructed, the "calorimeter."[74] From the outset, Lavoisier and Laplace intended that their machine should be put to use in the investigation of affinities.[75] By then reifying the cause of heat and repulsion as caloric, Lavoisier offered the possibility of rooting the study of affinities in one of "revolutionary" chemistry's simple elements.[76]

Traditional affinity tables had been ordered according to individual sets of relations. Now, in place of a situation in which affinity tables were

72. *Elements of Chemistry*, xxi.

73. "Mémoire sur les nouveaux caractères à employer en chimie," *Méthode*, unpaginated in book.

74. In the *Mémoire sur la chaleur* (Paris, 1783), their machine was called just that. Only in the *Traité* of 1789 did Lavoisier name it "calorimeter." See *Elements of Chemistry*, 343–356.

75. "The equilibrium between the heat, which tends to separate the particles of bodies, and their mutual affinities, which tend to bring them together, can furnish a very precise means of comparing these affinities with each other." *Memoir on Heat,* trans. Henry Guerlac (New York: Neale Watson Academic Publications, 1982), 28.

76. In the *Elements of Chemistry* Lavoisier added this rider to his definition of caloric: "We are not obliged to suppose this to be a real substance, it being sufficient . . . that it be considered as the repulsive cause, whatever that may be" (p. 5).

becoming increasingly complex and cumbersome through a proliferation of data and instrumental qualifications,[77] caloric offered a preorganized field of research. The laws of affinity might be determined by instrumentally comparing their forces with the repulsive force of caloric. But to be made visible, caloric's own activity had to be instrumentally bound, its causal presence marked and stabilized in the evidential context of Lavoisier's overall system.[78]

Caloric's release, sparked by a given process of de- or recomposition in the calorimeter's innermost compartment, melted a measurable quantity of ice in its adjacent chamber. Caloric's otherwise invisible presence was thus marked on one side by the progress of a chemical operation (its nature determined by the chemical investigator), and on the other by ice's measurable change of state. Once encased in the calorimeter (both instrumentally and nominatively), caloric provided a sort of causal glue, incorporating all of nature's processes into "revolutionary" chemistry's space of enquiry and manipulation.

Although not the subject of tabular presentation in its own right, caloric stood as the instrumentalized catalyst of all chemical activity. Now, not only could individual affinities be investigated and recorded by laboratory means; they were fundamentally subject to manipulation, themselves fully instrumentalized. As evidence for this, we need only turn to Lavoisier's introductory explanation of his "table of binary combinations of oxygen with simple substances." As portrayed in all his experimental descriptions associated with the table,[79] Lavoisier deemed caloric's introduction a necessary prerequisite for the oxygenation of a body to ensue:

> Nature . . . may be assisted by art, as we have it in our power to diminish the attraction of the particle of bodies almost at will by heating them, or, *in other words,* by introducing caloric into the interstices between their particles.[80]

Fire had been the steadfast companion and tool of chemical artisans since time immemorial. But now the cause of heat and rarefaction was

77. It was generally recognized that to keep up with the growing quantity and complexity of data, affinity tables would have to become increasingly complex. Taking the point to its extreme, Lavoisier wrote, "But, to obtain tables which shall strictly agree with experiment, it is necessary, as it were, to form a table for each degree of the thermometer." Quoted in Richard Kirwan, *An Essay on Phlogiston* (London, 1789), 48.

78. On the question of negotiating experimental closure in terms of what instruments record, see Trevor Pinch, "Toward an Analysis of Scientific Observation: The Externality and Evidential Significance of Observational Reports in Physics," *Social Studies of Science* 15 (1985): 3–36.

79. *Elements of Chemistry*, 53–82.

80. Ibid., 185–186. My emphasis.

reified as one of the chemical world's most fundamental elements. Caloric was firmly situated—both instrumentally and theoretically—as the constant and necessary instigator of all chemical activity, known only through the effects to which it could be made and measured to give rise. If it pointed back to anything, it was to the very nature of chemistry's scientific enterprise. Rather than consisting of a set of practices and catalogs intended to guide chemists through the unknown weave of nature, one thread at a time, chemistry was now a disciplined field of study. Its disciplinarity came not so much from the particularity of its theoretical positions and practices as from their overall structure. As represented especially in the *Traité*'s tables of binary combinations, chemical science stood as the circumscribing domain of both manipulative practice and the purported activity of nature itself. Indeed, the two had been conflated and rendered continuous to the point that there could be no more separating them.

Conclusion

Part II of Lavoisier's *Traité élémentaire de chymie* has been described as "an extended affinity table with commentary."[81] It is the purpose of this essay to argue the contrary. To characterize Lavoisier's systematic presentation simply as the culmination of a continuous development is to ignore the structural transformation that occurred in the discourse and disciplinary practice of eighteenth-century chemistry. In other words, it is to ignore the very heart of what constituted the Chemical Revolution.

The preceding hermeneutic reading of eighteenth-century chemistry's tables has revealed the following. By analyzing these tables in terms of their distinct structures of presentation, rather than narratively as commensurable sets of facts, we have observed a fundamental shift in chemists' construction and projection of both their discipline and their relation to nature. For most of the eighteenth century, chemists sought to discipline their practice and the world as a cooperative venture, the results of which they enunciated in terms of a theoretically neutral accrual of matters of fact. It was thought that by uncovering as many individual relations between (albeit compound) chemically active substances (relations of elective attraction and/or solute relations) and organizing them into synoptic tables, chemists could literally—and quite coercively—induce nature to reveal the truth of its own structure. Wrangling over the merits of competing theoret-

81. A. M. Duncan, "The Functions of Affinity Tables and Lavoisier's List of Elements," *Ambix* 17 (1970): 28–42. (See 41.)

ical systems of explanation, in the interim, would serve only to derail the communal enterprise.

As the quantity of matters of fact mounted, however, and the complexity of their manipulatively revealed characteristics increased, the chemical world picture grew ever more subject to contention. This was necessarily so because of the very nature of chemistry's endeavor. By subjecting the world to laboratory manipulation, chemists had always enforced some manner of direction on "nature's revelation."[82] The failure or refusal to recognize the organizing activity inherent in (increasingly standardized) laboratory manipulation precipitated a crisis within the fledgling discipline, precisely because this lack of recognition was formally embodied in chemistry's language of articulation. Discursively clinging to a position of theoretical neutrality meant that as manipulative practices and the phenomena they revealed grew increasingly complex, the crisis deepened. Chemists had no means of mediating among themselves and their findings. No "natural" structure with which to order the world of revealed relations presented itself, and chemists were disciplined against providing one of their own making.

Such was the situation until the appearance of the *Méthode de nomenclature chymique* and Lavoisier's *Traité élémentaire de chymie*. As embodied in the language and tables of these works, chemistry was now disciplined by a discursive structure that projected itself from the start to organize the world through the power of its all-encompassing, determinative reach. Henceforth, when chemists spoke of nature, it was with a language borne explicitly of laboratory production.

I would like to thank the Huntington Library research staff for their kind assistance, and my colleagues at Wetenschapsdynamica—especially Olga Amsterdamska, Loet Leydesdorff, and Stuart Blume—as well as Peter Dear, David Gooding, Greg Myers, and Trevor Pinch for their support, encouragement, and editorial comments.

82. As the phenomenology of technics has made clear, instrumental intervention—owing to the non-neutral essence of any and all tools—always directs us simultaneously toward certain perceptions and abilities and away from others. For a full discussion, see Donald Ihde, *Technics and Praxis*, Robert Cohen and Marx Wartofsky, eds., *Boston Studies in the Philosophy of Science* vol. XXIV (Dordrecht: D. Reidel, 1979).

Part III

Textual Generation of Knowledge

Peter Dear

5. Narratives, Anecdotes, and Experiments: Turning Experience into Science in the Seventeenth Century

Introduction

HISTORIANS ROUTINELY REFER to Baconianism, the Royal Society, and the metaphor of reading the book of nature (with its Paracelsian as well as Galilean overtones) to argue that the seventeenth century saw a move towards discovering nature through the senses, using observation and experiment. Certainly, something happened to experience in the seventeenth century; talk of experimental and experiential demonstration, or sometimes "ocular" demonstration—culminating in the term "experimental philosophy"—rapidly became widespread. But discussion of the theme has long tended to focus on the elucidation of Koyréan metaphysical constructs to legitimate and give sense to empiricism. The result has been a neglect of actual scientific practice.[1] Ironically, perhaps, the present study itself seeks to isolate an aspect of empiricism that has no direct relation to what Galileo actually did with inclined planes in the privacy of his own home, or what Boyle did with his furnace or his air pump.[2] It concerns scientific *literary* practice, regarded as a crucial feature of scientific practice as a whole, and it contends that an account of an experiment is an essential part of its performance.

1. On "post-Koyréan" intellectualist history of science of the period, see Steven Shapin, "Social Uses of Science," in G. S. Rousseau and Roy Porter, eds., *The Ferment of Knowledge: Studies in the Historiography of Eighteenth-Century Science* (Cambridge: Cambridge University Press, 1980), 93–139, esp. 105–111. Even apparently notable exceptions such as the account in the "Harvard Case Histories in Experimental Science" of pneumatic experiments allow experiment to play only an unproblematic directive role in the development of theoretical ideas. The important recent study that redirects attention toward experiment itself is Steven Shapin and Simon Schaffer, *Leviathan and the Air-Pump: Hobbes, Boyle and the Experimental Life* (Princeton, N.J.: Princeton University Press, 1985).

2. On the latter, see Steven Shapin, "The House of Experiment in Seventeenth-Century England," *Isis* 79 (1988): 373–404.

The old controversy over whether Galileo actually carried out the experiments, or contrived experiences, that he describes in his writings has now more or less abated. The consensus these days, based on study both of his manuscript work sheets and of his methodological practices, is that he did develop a lot of his ideas in concert with physical apparatus, although according to what precise relationship remains unclear.[3] Nonetheless, his publications are still recognized to contain, alongside accounts of experiences that Galileo probably tried for himself, a good number of "thought experiments." Those things usually considered his "real experiments" and those usually considered his "thought experiments," however, differ less than is often supposed. They share certain characteristics that are embodied in a very basic, constitutive feature of Galileo's work: the way in which experience appears as part of his finished argumentation. In a sense, all of Galileo's experiments are "thought experiments," or would have been if he could have managed it.

Looking at the way in which Galileo talked about manipulation of apparatus and its experienced behavior is crucial to understanding the construction of his science. It is, perhaps, tempting to describe his literary practice as simple packaging, or form, or "mere rhetoric"; to assert that the important questions relate to what he was really doing, not to how he presented his activities to the public. That would be wrong, however: the implied notion of what Galileo was "really doing" turns out to be strictly incoherent. What does it mean to *do an experiment*? Can we legitimately say that rolling balls down inclined planes is really "doing an experiment" even if the construal of those actions by their performer is quite different from the one that later experimentalism of a characteristically modern kind would give it? We cannot, because of what it takes to ascribe meaning to actions.

An account of an action is an inseparable part of its meaning, just as the meaning of the account itself relies on its implicit referent. Thus the meaning of an account of an experimental event—that which makes it an account of an experimental event rather than a series of marks on paper—is provided by its implicit reference to a spatiotemporally defined region of clinking glassware or grooved pieces of wood being manipulated by a human agent. The meaning of that spatiotemporal region itself—what makes it discernable as an experimental event—is conferred, reciprocally,

3. For extensive references, see R. H. Naylor, "Galileo's Experimental Discourse," in David Gooding, Trevor Pinch, and Simon Schaffer, eds., *The Uses of Experiment: Studies in the Natural Sciences* (Cambridge: Cambridge University Press, 1989), 117–134; see also references in n. 11 below.

by the *account* of an experimental event. In other words, there cannot be an account of an experimental event without reference to the spatiotemporally defined region, while the spatiotemporally defined region cannot *be* an experimental event without its constitution as such in the account. Looking at the account, the presentation of the formal establishment of knowledge-claims, is therefore inseparable from, and in fact crucial to, an understanding of what is happening epistemologically.[4]

An account of an experiment, in turn, is a piece of social currency. As Owen Hannaway has noted in regard to Andreas Libavius's view of the correct organization of chemistry as a public endeavor rather than a private, alchemical practice, the experience of an individual *per se* cannot be expressed in language, because language is essentially and irredeemably social. The very act of verbalizing experience implies the public constitution of knowledge. Thus Libavius's task "was to capture the verbal units of collective experience which flitted between nature and text, to pin these down by definition, and to order them neatly on the page to produce a methodized art."[5] The following section will show that the textual functions and effects implied by such a view of natural knowledge encompassed a broad sweep of philosophical activities in the seventeenth century stretching well beyond chemistry, because epistemology, method, and text were woven together in the assumptions of that dominant scholastic pedagogy which took Aristotle as its touchstone of legitimacy.

Galileo, Fabri, and the Philosophical Scope of Experience

The following pages employ the term "experiment" in a special way. An "experiment" will be a historical event in which an investigator experiences

4. A methodological dictum of Harold Garfinkel is precisely to the point here: "members' accounts, of every sort, in all their logical modes, with all of their uses, and for every method for their assembly are constituent features of the settings they make observable." Garfinkel, *Studies in Ethnomethodology* (Cambridge: Polity Press, 1984), 8. For a penetrating consideration of similar questions concerning geological fieldwork rather than experiment, see Martin J. S. Rudwick, *The Great Devonian Controversy: The Shaping of Scientific Knowledge Among Gentlemanly Specialists* (Chicago: University of Chicago Press, 1985), 429–435. For a recent discussion of related issues, focusing on the period under review here, see Geoffrey Cantor, "The Rhetoric of Experiment," in Gooding et al., *The Uses of Experiment,* 159–180. See also Bruno Latour, *Science in Action: How to Follow Scientists and Engineers Through Society* (Cambridge, Mass.: Harvard University Press, 1987), chap. 1, and other references in notes 5 and 6 to "Introduction," this volume.

5. Owen Hannaway, *The Chemists and the Word: The Didactic Origins of Chemistry* (Baltimore and London: Johns Hopkins University Press, 1975), 146. Hannaway stresses the dimension of Lutheran "civic humanism" informing Libavius's schoolmasterly approach to learning.

the behavior of a contrived setup, or apparatus, and uses, or might use, a report of that historical event as an element in constructing an argument intended to establish or promote a knowledge-claim. An experiment, therefore, is only an experiment if it appears as one in scientific discourse, or might well do so given the context in which it was created.

Thus, for example, when Galileo rolled balls down inclined planes, and shot them off the edges of tables to measure their distances of travel, he was not performing experiments unless he did these things as underpinning for their formal presentation in his writings as discrete historical events. But such is not the case. His famous description of inclined planes, adduced to justify his law of fall in the *Discourses and Demonstrations on Two New Sciences* of 1638, now appears more problematic than was recognized in the days of Koyréan and anti-Koyréan jousting over Galileo's credentials as an experimental scientist. Instead of describing a specific experiment or set of experiments carried out at a particular time, together with a detailed quantitative record of the outcomes, Galileo just says that with apparatus of a certain sort, he found the results to agree exactly with his theoretical assumptions—having, he says, repeated the trials "a full hundred times." He had shortly before claimed to have done this "often."[6] Both phrases are just ways of saying "again and again as much as you like": Galileo is establishing the authenticity of the experience that falling bodies do behave as he asserts by basing it on the memory of *many instances*—a multiplicity of unspecified instances adding up to experiential conviction. This is a technique that contemporary scholastic writers, preeminently Jesuit writers in the applied mathematical sciences, used routinely, and explicit justification of frequent repetition of trials appears in a 1613 work on optics by the Antwerp-based Jesuit Franciscus Aguilonius.[7]

At one point in the treatise, Aguilonius discusses the means of establishing the principles or axioms at the base of a formal science. Following Aristotle, he understands scientific knowledge as conforming to an axiomatic deductive structure analogous to that of geometry, Aristotle's apparent model. Deductive inferences are to be made on the basis of universal statements that act as premises, much as theorems in Euclidean geometry

6. *Le Opere di Galileo Galilei*, Edizione Nazionale, ed. Antonio Favaro, vol. VIII (Firenze: G. Barbèra, 1898): "per esperienze ben cento volte replicate. . . , " 213; "molte volte," 212; trans. Stillman Drake in Galileo Galilei, *Two New Sciences* (Madison: University of Wisconsin Press, 1974), 170, 169.

7. For full discussion, see Peter Dear, "Jesuit Mathematical Science and the Reconstitution of Experience in the Early Seventeenth Century," *Studies in History and Philosophy of Science* 18 (1987): 133–175.

are demonstrated from fundamental definitions and axioms. The central requirement is that the truth of the fundamental statements, or principles, from which a science is developed be immediately evident. An Aristotelian science does not involve formal procedures for demonstrating its own principles; the logic of the deductive structure requires that they simply be accepted at the outset. In geometry, the ideal is intuitive obviousness, as with the Euclidean (and Aristotelian) example that "the whole is greater than its part." Sciences dealing with the physical world, however, need empirical principles, which by their very nature cannot be intuitively obvious; they must be rendered acceptable through appeal to experience. Aguilonius considers how this must be done:

> A single sensory act does not greatly aid in the establishment of sciences and the settlement of common notions, since error can exist which lies hidden for a single act; but having been repeated time and again, [the act] strengthens the judgement of truth until [that judgement] finally passes into common assent; whence afterwards they [i.e., the "common notions"] are put together, through reasoning, as with the first principles of a science.[8]

This justification of multiple repetition has nothing to do with epistemological problems of induction; it concerns simply the avoidance of deception by the senses or by choice of an atypical instance, so as to ensure a reliable report of how nature actually behaves "always or for the most part," as Aristotle put it.[9] The relevant aspects of nature are themselves neither opaque nor capricious—if they were, there could be no question of making a science of them by having their characterizations pass into "common assent." Aguilonius's remarks tacitly appeal to Aristotle's definition of "experience" in the *Posterior Analytics:* "from perception there comes memory . . . and from memory (when it occurs often in connection with the same thing), experience; for memories that are many in number form a single experience."[10]

Recent research has shown that Galileo aimed at developing scientific

8. "Non enim ad scientiarum primordia, communiumque notionum constitutionem, unicus actus magnopere iuvat; siquidem error huic subesse potest, qui lateat, at saepè ac saepiùs repetitus iudicium veritatis corroborat, quousque tandem in communem assensum transeat. unde posteà velut ex primis principiis scientiae per ratiocinationem colliguntur." Franciscus Aguilonius, *Opticorum libri sex* (Antwerp, 1613), 215–216. (All translations throughout are my own unless otherwise noted.)

9. *Metaphysics* VI.2.

10. *Posterior Analytics* II.19, trans. Jonathan Barnes in Barnes, ed., *The Complete Works of Aristotle: The Revised Oxford Translation* vol. I (Princeton, N.J.: Princeton University Press, 1984).

knowledge, whether of moving bodies or of the motion of the earth, according to an Aristotelian (or Archimedean) deductive formal structure; in other words, according to the dominant model of the time. His literary practice regarding experience bears out that finding.[11] Galileo is a long way from the literary construction of an experiment in the sense of a reported, singular historical event. Galileo did not say, "I did this and this, and this is what happened, from which we can conclude . . ." and so forth; instead, he wrote, "this is what *happens*." The effect of naturalness attaching to this form of assertion is what renders Galileo's use of experience tantamount to the invocation of thought experiments: the reader is reassured that the world's working in a particular way is entirely to be expected, entirely consonant with ordinary events. There was nothing contentious or novel about such a construal of experience; Galileo could even allow it to be exhibited in the earlier *Dialogue* by Simplicio, his Aristotelian straw man, with no sense of danger to himself. In the course of the famous exchange concerning the dropping of weights from the mast of a moving ship, Salviati asserts that the outcome can be known without resort to experience. Simplicio retorts incredulously: "So you have not made a hundred tests, or even one? And yet you so freely declare it to be certain?"[12] The Galilean figure of "a hundred times" thus appears even in the construal of proper scientific experience by a notional opponent. For Galileo, the proper construal of experience was not an issue.

We can now see why Galileo's literary presentation of experience in the inclined plane example is sharply distinguished from the strict deductive structure constituting his formal science of motion. The material comes in the Third Day of the *Discourses*, and the science of motion proper is constituted by a deductive geometrical treatise in Latin. The presentation of the Latin treatise is interrupted periodically by discussion, in Italian,

---

11. Outstanding studies of Galileo's methodological ideals and practice include Winifred L. Wisan, "Galileo's Scientific Method: A Reexamination," in Robert E. Butts and Joseph C. Pitt, eds., *New Perspectives on Galileo* (Dordrecht: D. Reidel, 1978), 1–57; Ernan McMullin, "The Conception of Science in Galileo's Work," in ibid., 209–257; William A. Wallace, *Galileo and His Sources: The Heritage of the Collegio Romano in Galileo's Science* (Princeton, N.J.: Princeton University Press, 1984). Although Charles Schmitt identified systematic differences in the handling of experience between Galileo's early writings and those of the Paduan Aristotelian Zabarella, those differences tend to vanish when comparing Galileo with Jesuit authors. Charles B. Schmitt, "Experience and Experiment: A Comparison of Zabarella's View with Galileo's in De motu," *Studies in the Renaissance* 16 (1969): 80–138.

12. "Che dunque voi non n'avete fatte cento, non che una prova, e l'affermate così francamente per sicura?" Galileo, *Opere*, vol. 7 (Firenze: G. Barbèra, 1897), 171; trans. Stillman Drake in Galileo Galilei, *Dialogue Concerning the Two Chief World Systems—Ptolemaic and Copernican*, 2d ed. (Berkeley: University of California Press, 1967), 145.

between the three participants in the dialogue, and it is in one of those interruptions that the justificatory account of experiential support for the assertions of the formal treatise appears. Galileo tried as much as he could to make the basic assumption of that treatise—namely, that falling bodies accelerate such that their speed increases in proportion to time elapsed—appear intuitively obvious, but the Italian gloss bears witness to his failure. It tries to bolster the treatise with an appeal to universal experience, but the constructed, and in fact therefore essentially unobvious, recondite, character of that experience is precisely the reason why it could not appear in the formal science. Galileo's problem was that a true science had to rely on evident and universally acceptable premises. In having to adduce contrived experiences that were not themselves obvious, Galileo was deviating from that requirement.[13]

In a philosophical textbook of 1646, the Jesuit theologian and natural philosopher Honoré Fabri had a number of things to say on these issues, with specific reference to Galileo's work on falling bodies.[14] His overriding concern was to deflate the pretensions of Galileo's "science of motion" to proper (Aristotelian) scientific status; to undermine the rhetorical strivings of Galileo's text. Using the language of methodology as his weapon, Fabri focuses on Galileo's use of experience: Galileo used experience in his vindication of a formal science in a way that could not fulfill the demands put on it.[15] Fabri uses the Latin word "experimentum" rather than "experientia," but the distance between his meaning and the sense nowadays attached to the word "experiment" in English—or the sense that the word "experiment" came to hold for the early Royal Society of London a little later in the century—is considerable. Fabri's use of *experimentum* is just a form of the Aristotelian concept of "experience."[16]

Fabri defines his terms carefully:

13. Galileo, *Opere*, vol. 8, 190–213; Drake, trans., *Two New Sciences*, 147–170.

14. The work appeared under the name of one of Fabri's students as being a version of Fabri's lectures at the Jesuit college at Lyons; it was apparently fully authorized and understood as such by contemporaries (see references in n. 25 below): *Philosophiae tomus primus: qui complectitur scientiarum methodum sex libris explicatam: Logicam analyticam, duodecim libris demonstratam, & aliquot controversias logicas, breviter disputatas. Auctore Petro Mosnerio Doctore Medico. Cuncta excerpta ex praelectionibus R. P. Hon. Fabry. Soc. Iesu. Lugduni, sumptibus Ioannis Champion, in foro Cambij. M.DC.XLVI.*

15. On the functions and characteristics of "method talk" in science, see John A. Schuster, "Methodologies as Mythic Structures: A Preface to the Future Historiography of Method," *Metascience: Annual Review of the Australasian Association for the History, Philosophy and Social Studies of Science* 1/2 (1984): 15–36; also essays in John A. Schuster and Richard R. Yeo, eds., *The Politics and Rhetoric of Scientific Method: Historical Studies* (Dordrecht: D. Reidel, 1986).

16. Contemporary terminology, it should be remembered, made no clear distinction between the words "experientia" and "experimentum," so that each will usually be translated in

A physical experience [*experimentum*] is the behavior of some sensible thing, physically certain and evident—that is, such that it cannot fail this side of a miracle.[17] For example, at one time I see a stone move, at another I see it not move; I see the same thing with a sphere of lead and of wood; I feel the greater blow of a stone falling from a greater height, etc.[18]

A little further on he gives additional examples, somewhat more substantive this time, of the sorts of things that "all experiences agree in"—such as projectile motion not lasting indefinitely, or the acceleration of the natural motion of freely falling bodies. These, he says, "are established from most certain experiences," meaning from reliable "common experience" of the Aristotelian kind.[19] Such experiences, of course, render immediately accessible the validity of the argument based on them; the reader is given no room to doubt them or to consider the trustworthiness of the author's assertion. These are just what everyone knows, and the acceptance of them therefore acts as a kind of condition for reading the text.

Fabri's expectations about what true scientific knowledge should accomplish were fundamentally the same as Galileo's: they were scholastic-

this chapter as "experience"; see Schmitt, "Experience and Experiment," esp. 86–92. The use of "experimentum" in reference to "experience" in the Aristotelian sense is found, for example, in Riccioli: "Viguit iam inde à viginti & amplius saeculis in Academiis Physicorum, praesertim Peripateticorum cum Aristotele I. de caelo cap.88. illud ab experimentorum inductione collectum principium & axioma, per se satis notum sensui: Gravia naturali motu descendentia per medium levius, eò velociùs ac velociùs continuè moveri, quò propiùs accedunt ad terminum, ad quem tendunt. . . ." *Almagestum novum. Astronomiam veterum notamque complectens observationibus aliorum, et propiis novisque theorematibus, problematibus, ac tabulis promota. . . . Auctore P. Ioanne Baptista Ricciolo, Societatis Iesu Ferrariensi* (Bononiae, 1651), Part 2, 381, col. 2. On the other hand, a work of 1648 by Iohannes Chrysostomus Magnenus on atomism allows the following use of "experientia," recalling Aguilonius's methodological remarks: "Experientias accurate factas tanquam principia per se nota admittere" (quoted in Christoph Meinel, "Early Seventeenth-Century Atomism: Theory, Epistemology, and the Insufficiency of Experiment," *Isis* 79 (1988): 68–103, on 80). For an apparent refinement of usage, making an effective distinction between "experimentum" and "experientia," by Christopher Scheiner earlier in the century, see Dear, "Jesuit Mathematical Science," 156–157; Scheiner's practice does not seem to have been usual, however, although it would be possible to read it into the passage from Riccioli just quoted. I shall translate "experientia" and "experimentum" indifferently throughout as "experience," while allowing the reader access in the notes to the original term.

17. "Physical certainty" took its place between metaphysical certainty (the highest grade) and moral certainty (the lowest grade) in contemporary classifications; the terms find their echoes in Descartes. For a Jesuit exposition of the matter, see Roderigo de Arriaga, *Cursus philosophicus* (Antwerp, 1632), 226, col. 1.

18. "Experimentum Physicum, est effectus aliquis sensibilis, certus & evidens Physicè, id est, ita ut citra miraculum fallere non possit, v.g. video lapidem modò moveri, modò non moveri; idem video in globo plumbeo, ligneo; sentio maiorem ictum lapidis ex maiori altitudine cadentis, &c." Fabri, *Philosophiae tomus*, 88, part of a chapter headed: "De Principiis & demonstrationibus Physicis."

19. Ibid., 88–89: "omnia experimenta consentiant"; "quae ex certissimis experimentis constant."

Aristotelian. Their legitimacy, or potential plausibility to a seventeenth-century reader, depended on an orthodox, conventional framework familiar to anyone who studied natural philosophy and logic in that period in Europe. Fabri's position rested on his view of the role that experience ought properly to play in constructing knowledge of nature. It can, perhaps, best be explicated by a contrast with the modern hypothetico-deductive view of scientific procedure. Some version of the latter, whether confirmationist or falsificationist, would place experience, at least as regards its formal justificatory role, at the *end* of a logical structure of deduction from an initial hypothesis: the hypothesis yields conclusions regarding observable behavior in the world, and experiment or observation then steps in to confirm or falsify these predictions—and hence, in a logically mediated way, to confirm or falsify the original hypothesis itself.[20] A methodological Aristotelian, however, approached these issues in a quite different fashion. Since the point of Aristotelian scientific demonstration was to derive conclusions deductively from premises that were already accepted as certain—as with those of Euclidean geometry—there was no question of testing the conclusions against experience. The proper place for experience was in grounding the inductive generalizations contained in the original premises, as Aguilonius assumed. Once they had been established, so too, in effect, had the conclusions deduced from them.[21]

Fabri uses a particular terminology to talk about the place of experience in demonstrative science, a terminology the precise details of which may not be commonplace, but which nonetheless relies on this fundamental construal of the issues. He calls the basic experiential statements that act as premises in the syllogistic demonstrations of natural philosophy "physical hypotheses." A physical hypothesis derived from experience is therefore a universal statement such as "falling bodies accelerate."[22] Fabri's use of the word "hypothesis" is not intended to refer to a statement or group of statements that are conjectural, awaiting test through the empirical investigation of their consequences. He uses it to mean "fundamental statement," a statement that is suitable to stand as a premise at the beginning of

20. Classic expositions of variants of this picture are Ernest Nagel, *The Structure of Science* (New York: Harcourt, Brace, 1961); Karl Popper, *The Logic of Scientific Discovery* (London: Hutchinson, 1959).

21. Cf. Dear, "Jesuit Mathematical Science."

22. Fabri, *Philosophiae tomus*, 88. The meaning of the term "hypothesis Physica" here seems to reside in the invocation of "physical certainty"; see n. 17, above. It does not seem to have been in common philosophical usage; its appearance much later in Stephanus Chauvin, *Lexicon philosophicum*, 2d ed., (Leeuwarden, 1713; facsimile reprint Düsseldorf: Stern-Verlag Janssen, 1967), 297, is Cartesian-influenced, and seems to be rather different.

logical demonstration. A genuine "physical hypothesis" is a physically certain principle founded on universal experience, the latter itself the product of many memories of the same thing. The label means something else as well, however: this designation of the principles appropriate for developing a science of motion includes the word "physical." In the scientific discourse in which Fabri (like Galileo) participates, "physical" means, among other things, "not mathematical."[23] Galileo had clearly regarded the present subject matter as mathematical, whereas Fabri regards it as physical. Fabri asserts that "no physical hypothesis"—such as Galileo's concerning uniform acceleration—"is to be sought from an experience that is not established with physical certitude."[24]

Fabri's arguments against Galileo show the practical import of these terms and definitions. He wants to contest Galileo's odd-number rule for falling bodies, the idea that the distances covered by a freely falling heavy body progress in the sequence 1, 3, 5, 7, and so on. He argues against that on specific causal and analytical grounds in a separate treatise,[25] but here he intends to show that claims to have established the rule from experience are false by the very nature of the thing: experience is incapable of demonstrating something of that sort. Those people are simply wrong, he says, who assert the odd-number progression on the grounds that it agrees with "the most demonstrative experience that the space acquired in the second period of time, equal to the first, is triple that acquired in the first period," and his central point is that the senses cannot judge those distances and times

23. For the institutional and intellectual importance to Jesuits of the disciplinary boundary represented (and maintained) by the distinction between "mathematical" and "natural philosophical" scientific genres, see Dear, "Jesuit Mathematical Science," sect. IV. "Mathematical" sciences, it should be remembered, included sciences of nature, such as geometrical optics, that utilized theorems drawn from arithmetic or geometry and that regarded only the *quantitative* properties of things; "physics," or "natural philosophy," addressed *qualities*. Although Aristotle's ideal of deductive scientific demonstration was modeled on Greek geometry, it was intended to apply to both mathematics and physics despite their distinct subject matters.

24. "Hinc nulla hypothesis Physica ab eo experimento petenda est, quod non est certum certitudine Physicâ" (Fabri, *Philosophiae tomus,* 88). A brief account of this discussion appears in David Clough Lukens, "An Aristotelian Response to Galileo: Honoré Fabri, S.J. (1608–1688) on the Causal Analysis of Motion" (Ph.D. diss., University of Toronto, 1979), 115–118.

25. The *Tractatus physicus de motu locali* of 1646, also bearing the name of Mousnier and often bound together with the *Philosophiae tomus.* See, on its doctrines of fall, Lukens, "An Aristotelian Response," chap. 4; Stillman Drake, "Free Fall from Albert of Saxony to Honoré Fabri," *Studies in History and Philosophy of Science* 5 (1975): 347–366; idem, "Impetus Theory and Quanta of Speed," *Physis* 16 (1974): 47–65; and for discussion of reactions by contemporaries, with further references, Peter Dear, *Mersenne and the Learning of the Schools* (Ithaca, N.Y.: Cornell University Press, 1988), 215–218. Fabri's position flows from his construal of the question as physical rather than mathematical, and involves focus on causal explanation utilizing the idea of impetus. Baliani adopted a similar position: see Serge Moscovici, *L'expérience du mouvement: Jean-Baptiste Baliani, disciple et critique de Galilée* (Paris: Hermann, 1967).

precisely.[26] He had already explained that "the name 'experimentum' ought to exclude all that which does not fall under the senses," and the examples he gave were the equality of two times, or the equality of two distances fallen. If one of those distances or times were one-thousandth part greater than the other, he says, the senses typically would be incapable of discerning it. The essential problem with Galileo's odd-number rule was that it could not be based on experience, or "experimenta," because sensory data could never provide sufficient precision to guarantee it. Questioning the veracity of Galileo's experiential claims themselves was not part of Fabri's purpose.[27]

According to Fabri, then, Galileo was doubly guilty of solecism. He had confused distinct genres of scientific argument, treating physics, characterized by talk of causes and "physical hypotheses," as if it were mathematics, characterized by talk of formal relationships among quantities and ratios. In addition, he had purported to establish a "physical hypothesis," the odd-number rule, on experiential grounds that were inadequate. But Fabri did not use his assertion that experience could not support Galileo's odd-number rule to imply that statements about the ratios observed in natural acceleration can never be established scientifically (in the proper Aristotelian sense of the word). Galileo's talk about rolling balls down inclined planes suggested that those ratios could be established directly from experience, and that was the focus of Fabri's criticism; Fabri nonetheless had his own ideas on the matter that avoided the difficulties found in Galileo's. He claimed that his own characterization of natural acceleration was based on a different methodological approach: a statement about such a thing can be validly established either from another hypothesis that is itself properly derived from experience, or from some other principle. The statement so established will not itself be a physical hypothesis; instead, it will be a theorem, *derived* from a physical principle. That is the status Fabri accords his own claim that the distances covered in successive equal periods of time by freely falling bodies follow the series of integers 1, 2, 3, 4, . . .[28]

Despite their differences, Galileo and Fabri concurred in allowing no functional place for the experiment, the set-piece event: one did not create

26. Those who would maintain that the odd-number progression "probatissimo constet experimento; in secundo tempore aequali primo, acquiri spatium triplum acquisitum in primo tempore" (Fabri, *Philosophiae tomus,* 88).

27. "Porrò experimenti nomine carere debet, id omne quod in sensum non cadit" (ibid., 88). Fabri, it should be noted, accepted the odd-number rule *phenomenologically.*

28. Lukens, "An Aristotelian Response," chap. 4; Drake, "Free Fall"; idem, "Impetus Theory."

knowledge of nature by recounting histories, whether in a mathematical or a natural-philosophical science. There is a less than sharp line, however, between these commonplaces about "experience" and a more modern understanding of "experiment" as a discrete, contrived, and reported event fit for use in establishing knowledge-claims about nature. The ambiguity of many seventeenth-century accounts of philosophically relevant experience is precisely what has obscured the very recognition of the issue. Experiment emerged textually in the seventeenth century through a gradual stretching of the conventions already considered, but it did not acquire full legitimacy in an unproblematic way. Indeed, the early Royal Society's proclamation of a philosophy built explicitly on experimental events attended a forceful rejection of scholastic Aristotelian traditions. The effort to declare difference should not, of course, be taken entirely at face value; no doubt the smaller the difference, the greater the work expended in maintaining its appearance. But even the gradual establishment of narratives of events constituted a major epistemological reorientation. The following section examines an instance of the exploitation of narrative in a controversy, and the subtle handling required to remain within the bounds of properly scientific discourse. It also bears witness to the fact that, in the disagreement between Fabri and Galileo, it was Fabri who acted as the philosophical renegade. His methodological objections to Galileo were not shared by other Jesuits, whereas Galileo's own behavior can be understood only in terms of those Aristotelian epistemological commonplaces in which they dealt.

Arriaga, Riccioli, and the Vulnerability of Experience

In 1632 Roderigo de Arriaga, a Jesuit located in Prague, discussed falling bodies in the first edition of his *Cursus philosophicus*. As the date indicates, Arriaga was not responding to Galileo (whose *Dialogue* appeared in the same year), but to the increasing currency of ideas shared by others. Under one of the subheadings to a set of "disputations" on Aristotle's *On Coming-to-be and Passing-away* (using the, by then, usual humanistic Latin title *De ortu et interitu* in place of the medieval translation *De generatione et corruptione*), Arriaga maintains the proposition, "All heavy bodies fall downwards by themselves equally." The peculiar nature of his task is set by the admission that the proposition flies in the face of received opinion:

Hitherto, nothing has been so generally accepted among philosophers and other men as that bodies that are heavier consequently descend faster to the earth; to doubt concerning which was once a heinous thing. Hence, again, it was eminently fixed in the minds of all that a stone moves faster at about the middle [of its flight] than in its initial motion.[29]

Arriaga's stress on the universality of these convictions serves not to establish their solidity—as might easily have been the case under other circumstances—but to dramatize his own dissent from them. The general opinion as Arriaga handles it, rather than being a weighty argumentative factor, is itself made into a kind of natural curiosity: "it is wonderful, inasmuch as this matter can be so easily tested by experience, that never in all the centuries have they tested it even once, as they would have discovered thereby the falsity of their opinion." Arriaga proceeds to accredit his assertions, providing the following context: "a few years ago some people doubted that two globes equal in shape and size fall equally from a certain distance even if the [lighter] one is unconnected to the heavier, and likewise for any other shape." But now, Arriaga says, "I not only find this in [bodies of] equal shape and size, but in [all] bodies whatever, however much unequal in form, shape or size, so that one dry piece of bread-crust of two inches fell equally speedily from a very high place as a stone that I was able only with difficulty to hold with my hands." What should be noted here is that Arriaga makes his general claim of experienced knowledge ("I find this . . .") and only subsequently adduces a specific example by way of illustration. Even then, the establishment of a kind of universality is implied for the specific bread/stone experience; he makes clear that he has tried this "not once or twice, but often," so that there is nothing peculiar even about this particular instance of the general phenomenon. The requisite universality is then warranted by an extension even beyond Arriaga's personal experience itself; not only has he tried the phenomenon often, but also "in the presence of many people." Thus it has become truly evident. Arriaga concludes:

The same [holds] even more for a small stone together with any rock: indeed, a thick piece of paper, compacted from six or seven others into a flat and

29. "Nihil hucusque inter Philosophos ac reliquos homines ita receptum, quàm corpora, quò sunt graviora, eò velociùs descendere ad terram, de quâ re dubitare, olim fuisset nefas. Hinc rursus altè omnium animis insitum fuit, velociùs lapidem circa centrum moveri, quàm in initio motus." Arriaga, *Cursus philosophicus* (1632), 582, col. 1.

round shape and very small, falls as fast as a big stone. What philosopher (if experience and eyes were not witnesses greater than any objection) could say this without being considered foolish? Perhaps the same would happen in many received opinions, if we could examine them by experience itself.[30]

In sum, therefore, Arriaga attempts to invest his claims with credibility by stressing the frequency with which he has experienced the relevant behavior, as well as its sharing with others. This is exactly what is found in Galileo's account of fall along inclined planes, where Salviati stresses both the frequency of trial and his own participation—this was no private garnering of experience, but its public establishment, rendering it fit for general proclamation. In more practical terms, this passage of Arriaga's is designed to create a conviction, or at least a willingness to lend credence, in readers who have certain expectations concerning how experience ought properly to be discussed so as to establish a knowledge-claim—expectations that Aguilonius's remarks served to warrant. The operative element of this knowledge-creation is not Arriaga and associates dropping pieces of bread, but the rhetorical efficacy of Arriaga's discussion; more particularly, however (since something similar can necessarily always be said of any scientific text), the rhetorical efficacy in this case is one resting on implicit appeals to the familiarity of a scholastic rhetoric of experience.

Mere utilization of the appropriate rhetoric was not, of course, in itself sufficient to guarantee universal acceptance of one's claims. An opponent could always deploy similar weapons in an attempt to undermine the work of the original argument. In his 1651 blockbuster *Almagestum novum*, a work of astronomy and cosmography, the Jesuit astronomer Giambattista Ric-

---

30. "Et mirandum est, quòd cùm res haec experientiâ tam facilè comprobari possit, numquam per tot saecula vel semel id experti sint, ut inde suae opinionis falsitatem deprehenderent. Dubitarunt [*sic*: dubitaverunt?] hac de re aliqui ante aliquot annos in duobus globis aequalis figurae & magnitudinis, qui ex quacumque distantiâ aequaliter cadunt, licet sit unus decuplo altero gravior; & idem est de qualibet aliâ figurâ: ego verò non solùm in aequali figurâ & in magnitudine, sed in quibuscumque corporibus, quantumvis formâ, figurâ, magnitudine inaequalibus, idem deprehendo, ita ut unum siccum corticem panis duorum digitorum, cum saxo quod manibus vix tenere poteram, ex alto valdè loco aequè citò cadere, non semel aut bis, sed saepè expertus coram multis sim. Idem à fortiori de exiguo lapillo cum quocumque saxo: imò chartam crassam, ex sex vel septem aliis compactam in figurâ planâ & rotundâ ac valdè exiguâ, tam citò cadere quàm magnum lapidem. Quis hoc Philosophorum diceret (si experientia & oculi non essent testes omni exceptione maiores) qui non putaretur stultus? Ita fortè in multis receptis opinionibus contingeret, si eas experientiâ ipsâ examinare possemus." Ibid. This kind of presentation should not be regarded as original with Arriaga: for late sixteenth-century examples, see William A. Wallace, "Traditional Natural Philosophy," in Charles B. Schmitt, Quentin Skinner, Eckhard Kessler, and Jill Kraye, eds., *The Cambridge History of Renaissance Philosophy* (Cambridge: Cambridge University Press, 1988), 201–235, on 221–223.

cioli denied Arriaga's assertion about falling bodies, which also meant denying the concurring views of others, including other Jesuits. Undaunted by the proposition's growing number of supporters (much as Arriaga was undaunted by the previous uniformity of its denial), Riccioli attempted to discredit the experiences on which they relied. He especially addressed the Jesuits Niccolò Cabeo and Arriaga. Cabeo had discussed the question in a commentary on Aristotle's *Meteorology* (1646), and Riccioli outlines his claims thus:

> he affirms most emphatically from his own, often repeated experiences, [that] if two balls are dropped at the same time from the same height, one of one ounce and the other of ten pounds or whatever greater weight, either both being of lead, or one lead and the other stone or wood; providing that the air is still, and that the lighter one is not of so small a weight that it is not forceful enough to overcome the resistance of the air, or tosses about in the breeze (like a feather or piece of paper), it will happen that both reach the ground at the same moment, and no difference can be detected by the senses in the fall: from which he infers the speed of all falling bodies to be intrinsically equal.[31]

Then comes Riccioli's counterattack. Cabeo has, quite appropriately, stressed that he tried these things often; Riccioli wishes to weaken that powerful commonplace by suggesting that Cabeo's resultant expertise was not, in fact, suited to settling the question at issue. "However," he says, "I do not know from what height he released those balls." This was to be a telling point. But its full effect is achieved through the most powerful stroke at Riccioli's command: to insert himself into Cabeo's account and replace Cabeo's narrative voice with his own. He means to say that one cannot tell *from Cabeo's own account* what the height of release was. His own account will be more reliable than Cabeo's; he explains that "when we were at Ferrara at the same time in 1634" he participated in these very experiences.

Riccioli proceeds to detail what he learned from the experiences on which Cabeo apparently based his argument:

31. ". . . asseverantissimè affirmat ex proprijs saepiusque iteratis experimentis, si globi duo simul dimittantur ex eadem altitudine, unus unciae unius, alter decem librarum, vel cuiuslibet maioris ponderis, sive ambo sint plumbei, sive unus plumbeus sit, alter vel lapidus vel ligneus; dummodo & aër sit tranquillus, & illud quod levius est, non sit tantulae gravitatis, ut non valens vincere resistentiam aëris aut aurae fluctuet in aëre, (cuiusmodi esse plumam vel chartam) fore ut ambo eodem momento ad terram perveniant, nullumque sensibile discrimen in casu notari possit: ex quo infert, omnium cadentium velocitatem per se aequalem esse." Riccioli, *Almagestum novum*, Pt. 2, Lib.IX, Sect.IV, cap.XVI, 382, col. 2. Cabeo's claims appeared in Lib.I, pp. 97, col. 1 to 98, col. 1 of his commentary (Rome, 1646).

besides wooden balls, we released stones of diverse weights from the tower of
our chapel of the Society of Jesus, for one a bronze basin, for the other a
wooden board, having been put by me underneath, so that from the different
sounds I would distinguish better which one reached the ground faster.

He remembers certainly "that I noticed that the heavier one reached [the
ground] a little bit more quickly." Now the matter of the altitude of the
drop comes into play, and Cabeo's authority is crushingly demoted: "But
because that difference [in time of fall] was tiny [*exiguum*]—for the tower
at that place, from which they were released, did not exceed eighty feet—
for that reason he [sc. Cabeo] was never able to be persuaded to admit any
inequality or difference in the descent."[32] That constitutes Riccioli's final,
and no doubt very effective, word on the worth of Cabeo's claims.

Arriaga presents Riccioli with a different problem, insofar as Riccioli
cannot claim any privileged authority in rewriting Arriaga's text. Instead,
he has to rely on exploiting certain features of the text itself. Riccioli
portrays Arriaga as the crucial case: he cites him and others[33] as asserting
that "any two heavy bodies, whether of the same or of different species [i.e.,
material composition], or bulk, or figure, however much the difference in
weight, descend from the same height in the same time, and fall down with
equal intrinsic speed" and goes on to say that "truly, the foundation of so
great an assertion is the experience of Arriaga." This move serves to emphasize that the other statements of the disputed phenomenon are merely
derivative of Arriaga and should not be seen as independent authorizations
of it. Once having established the point, Riccioli proceeds to weaken
Arriaga's crucial experience,

by which he says that he has often released from the same table simultaneously a piece of dry bread crust of two inches in size, and a pen, with which

32. "Nescio autem ex quanta altitudine globos illos dimiserit, hoc tamen certò memini,
cùm essemus simul Ferrariae Anno 1634. & ex turri nostri templi Societatis IESU demit-
teremus lapides diversae gravitatis, nec non globos ligneos, me subiecta uni pelui aenea, alteri
tabula lignea, ut ex diversitate sonitûs meliùs distinguerem, uter citiùs ad terram perveniret,
advertisse id quod gravius erat aliquantulò citiùs pervenire. Sed quia illud discrimen exiguum
erat, neque enim turris ille locus, ex quo demittebantur, excedebat 80. pedes, idcirco ille
nunquam adduci potuit, ut eam vel ullam inaequalitatem admitteret, aut discrimen in lapsu."
Ibid. Note how Riccioli demotes Cabeo's claims by shifting them from statements about the
world to statements about Cabeo's own stubbornness: see the discussion of "modalities" in
Latour, *Science in Action*, 22–26 (an idea first presented in Latour and Steve Woolgar, *Laboratory Life: The [Social] Construction of Scientific Facts* [Beverly Hills, Calif., and London: Sage,
1979; 2d ed., Princeton, N.J.: Princeton University Press, 1986]).
33. Bartholomaeus Mastrius and Bonaventura Bellutus, in coauthored disputations on
Aristotle's *De caelo* and *Meteorologica* (Venice, 1640).

he wrote, and a great stone, which he could only with difficulty hold in his hands, and he has noticed these simultaneously to strike the pavement at the same moment, and to move equally fast, from which he exclaims and bemoans that in so easy a thing none of the philosophers has ever established [the characteristics of] descent by experience, but almost all have persisted in relying on their predecessors.[34]

Arriaga, of course, had claimed to release objects "from a very great height"; Riccioli, however, taking advantage of Arriaga's lack of specificity on the matter, uses the presence of bread crust in the account to reduce the dignity of Arriaga's experience to a matter of dropping things from a table (*ex mensa*), as if it were a matter of casual observation under inadequate circumstances. Rubbing salt into the newly created wound, Riccioli's final presentation of Arriaga's criticism of "the philosophers" contrasts the poverty of Arriaga's experience with his arrogance in assailing others.

Some years later, in the fifth edition of his *Cursus philosophicus*, Arriaga responded to unnamed critics. Specifying with nice imprecision the appropriate restriction of his claims to heavy bodies that fall directly down, rather than float in the manner of paper or feathers, he reasserts the "truth of experience" which obviates any need for "profound speculation." The recent writers who oppose his conclusion maintain the clarity of the "contrary experience" out of bitterness, and will not accept his refutations gracefully. Arriaga professes astonishment that others, judging the matter under the same circumstances, could deny "so many and such clear experiences in my favor"; nonetheless, so fair is he, that in the face of the "exclamations" against him, he says, "I have made the same experience again very often, even with some of them themselves in person, and just as I have taught, I have perceived the thing to be so."[35] These protestations,

34. ". . . asserunt, duo quaecumque corpora gravia sive eiusdem, sive diversae speciei, aut molis, aut figurae, quantumcumque differentia in gravitate, eodem tempore ex eâdem altitudine descendere, & aequali per se velocitate labi: Tantae verò assertionis fundamentum est experimentum Arriagae, quo ait se saepius ex mensa eâdem dimisisse simul siccum panis corticem duorum digitorum, & calamum, quo scribebat, & saxum ingens, quod vix manibus sustinere poterat, & advertisse haec simul eodem momento pavimentum percutere, atque aequè velociter moveri, ex quo exclamat & conqueritur, in re tam facili nullum unquam philosophorum experientia hunc descensum comprobasse, sed omnes ferè in fide parentum permansisse." Riccioli, *Almagestum novum*, Pt. 2, 382, col. 2. The final remark carries echoes of Galileo's caricatures of doctrinaire Aristotelians. Mastrius and Bellutus discuss Arriaga's work on pp. 175–179 of their treatise.

35. Arriaga, *Cursus philosophicus*, 5th ed. (1669), 691, original passage quoted in Charles B. Schmitt, "Galileo and the Seventeenth-Century Text-Book Tradition," in Paolo Galluzzi, ed., *Novità celesti e crisi del sapere: Atti del Convegno Internazionale di Studi Galileiani* (Firenze: Giunti Barbèra, 1984), 217–228, on 224n.; my translations. Arriaga's 4th ed. (Paris, 1647) is identical to the 1st on this material.

however, are now backed up by historical descriptions of specific trials, introduced in this way:

> Recently, from the top of the cupola of the Prague Cathedral, which is very high, I released from my hands at the same time a small stone and another more than twenty times heavier, and still both touched ground at precisely the same time. The same thing was done from Karlstejn Castle which is even higher, and they fell in exactly the same way, even if bodies of unequal weight were dropped.[36]

Riccioli, a prominent "recent writer" who denied Arriaga's claim, had ridiculed Arriaga's justification of it by exploiting the imprecisions and vaguenesses of his account. The fifth edition, with reports of actual trials and their locations, moved to fill the rhetorical breach.

Riccioli closed the section of his discussion that reviewed work by previous authors on falling bodies by describing the contrasting nature of his own subsequent presentation. "Hitherto," he says, "I have not spoken about the *opinions* of others, but about their *errors*, because the opposing of them by us and by others who have been present as witnesses to our observations, shortly to be reported, is indeed not just probable, but evident and certain." Riccioli's assertive confidence is unbounded: "and therefore," he continues, "we will proceed from the following experiences, not by way of likely conjectures but according to infallible physico-mathematical science, to certain conclusions."[37] Cabeo, Arriaga, and the rest were unreliable because they were incompetent; either their trials were misconceived or they reported them inaccurately. Somehow, by contrast, Riccioli's accounts of experiences would be established for the reader as adequate to the burden of proof they had to bear, and as entirely reliable.

---

36. Ibid., 224 and n.; trans. Schmitt.

37. "Hactenus de aliorum non dico opinionibus, sed erroribus, quia illarum oppositum nobis & alijs, qui testes adfuerunt observationibus nostris, mox referendis, iam non est probabile tantùm, sed evidens ac certum, idèoque non ex verisimilibus coniecturis, sed ex infallibili scientia Physico-Mathematica ab experimentis sequentibus ad conclusiones certas procedemus." Riccoli, *Almagestum novum*, Pt. 2, 383, col. 1. In light of the remarks in the first part of this chapter concerning the disciplinary distinction between physics and mathematics, Riccioli's term "physico-mathematics" deserves notice. As far as I know, it is a seventeenth-century neologism that comes into quite general use around the middle of the century among Jesuit mathematical scientists; for example, Riccioli's collaborator Francisco Maria Grimaldi's treatise announcing the diffraction of light is entitled *Physico-mathesis de lumine, coloribus, et iride* (Bologna, 1665). The word was a favorite of Descartes's early mentor Isaac Beeckman, and was used by others such as Mersenne, during the first half of the century (e.g. Mersenne's *Cogitata physico-mathematica* of 1644). Its adoption by Jesuits is particularly interesting because of its implied challenge to the existing curricular and disciplinary structure within which they worked; the issue is one well worth further investigation.

Experiences and the Establishment of "Expertise"

Riccioli's uses of the dominant scholastic conventions that constituted scientific experience throughout most of the seventeenth century reveal especially starkly their ambiguities and opportunities. His handling of the work of opponents involved attempts at investing himself with superior authority, derived from a purportedly greater understanding of how to make legitimate scientific experience. In other words, Riccioli tried to convince his readers that he possessed the right kind of *expertise*. The possession of experience in the appropriate Aristotelian sense was the foundation of a demonstrative science of nature. The extension of that experience to others occurred through the medium of the text whereby the author's expertise received its public warrant.[38] Riccioli's discussions of the behavior of pendulums and falling bodies in his *Almagestum novum* serve as useful illustrations of the techniques available for accomplishing this effect.

Since it was a work not just of astronomy but also more generally of cosmography, Riccioli devotes part of the *Almagestum novum* to the "sphere of the elements," the terrestrial realm; two of its chapters deal with the behavior of pendulums and falling bodies.[39] The first may be seen as a prolegomenon to the second insofar as its goal seems to be the establishment of instrumental techniques rather than the discovery of new aspects of nature. Later in the work, as a prologue to a discussion of the motion of the earth, there is another, much fuller study of fall.[40] The chapter on pen-

38. "Expertus" to mean "experienced" is classical Latin usage, although "expert" as a noun, like "expertise," is apparently recent usage in both English and French (the *Oxford English Dictionary*'s earliest example of the noun "expert" in English is from 1825). The issue of "expertise" has recently been considered in Rose-Mary Sargent, "Scientific Experiment and Legal Expertise: The Way of Experience in Seventeenth-Century England," *Studies in History and Philosophy of Science* 20 (1989): 19–45. However, while stressing (p. 28) the nature of "experience" in the English common-law tradition as a matter of "expertise" on the part of judge and jury, in a sense quite similar to that used here, Sargent never shows how this is supposed to apply to English experimental philosophy. It certainly stands a better chance of working in understanding Francis Bacon than Robert Boyle, Sargent's chief subject of examination: cf. Julian Martin, "'Knowledge is Power': Francis Bacon, the State, and the Reform of Natural Philosophy" (Ph.D. diss., University of Cambridge, 1988), esp. 247–255.

39. Riccioli, *Almagestum novum*, Part 1, "Liber secundus: De sphaera elementari," cap.XX, XXI. Some of this material is employed, in concert with relevant passages from Part 2, in Alexandre Koyré, "An Experiment in Measurement," in Koyré, *Metaphysics and Measurement* (Cambridge, Mass.: Harvard University Press, 1968), 89–117, on 102–108. Koyré's account is not always reliable and sometimes combines different passages in questionable ways.

40. On Riccioli's discussion of the earth's motion, see Alexandre Koyré, "A Documentary History of the Problem of Fall from Kepler to Newton: De motu gravium naturaliter cadentium in hypothesi terrae motae," *Transactions of the American Philosophical Society* n.s. 45 (1955), Part 4, 349–354; Edward Grant, "In Defense of the Earth's Centrality and Immobility: Scholastic Reaction to Copernicanism in the Seventeenth Century," ibid., n.s. 74 (1984), Pt. 4,

dulums is headed, "Concerning the oscillations of a pendulum suitable to measuring other motions and times, as much in elementary and compound bodies as in stars."[41] Riccioli nonetheless signals the comparative novelty of the enterprise, presenting the "Occasion for exploring the oscillations of the pendulum":

> When I was professor of philosophy at Parma, and wished to learn by what increment of velocity heavy bodies descend faster and faster towards their end, and saw that it required a very precise measure of time, and usually of its smallest sensible parts; the occasion at length offered itself to me to measure it.[42]

He found that other people who had written about it (including Galileo) had not actually measured the periods of oscillations. That provides the pretext for a review of what he has found, based on his own work as well as that of colleagues. In other words, he is not about to record specific experimental trials, but the distilled wisdom of his own experience of the behavior of pendulums. Accordingly, he presents a diagram of a pendulum apparatus together with its description and definitions of terms. The description is an abstract account of a *type* of apparatus rather than a report of an actual piece of equipment, taking the form of statements such as, "Let in the present diagram AB be an iron rod. . . ."[43]

Riccioli continues with a number of "propositions" mimicking geometrical presentation, the confirmation of which consists of statements of "experiences" (*experimenta*).[44] Again, these "experiences" are not given as reports of discrete trials. Thus the first proposition, stating the isochrony of pendulums, is justified first of all by a set of instructions. "Make yourself, then," says Riccioli, "a sundial from beaten tin [? *stanneo breve*], count the hours of one semiquadrant, or of ten minutes, so that there is less tedium at the beginning of the experiences . . ." and so on, concluding: "If, however, you were not diligent in doing this, there might be a difference of one or two oscillations between the vibrations of the first and second [periods of]

51–54; John L. Russell, S.J., "Catholic Astronomers and the Copernican System after the Condemnation of Galileo," *Annals of Science* 46 (1989): 365–386.

41. "De Perpendiculi Oscillationibus ad motus alios & tempora mensuranda idoneis, tam in Elementis & mixtis, quàm in syderibus." Riccioli, *Almagestum novum*, Pt. I, 84, col. I.

42. "Cum profiterer Parmae Philosophiam publicè, optaremque experiri quo incremento velocitatis Gravia descenderent velociùs, ac velociùs versus finem, videremque ad id requiri mensuram temporis subtilissimam, & minimarum fermè particularum eius sensibilium; Oblata est mihi tandem occasio huius mensurae. . . ." Ibid.

43. "Sit in praesenti diagrammate Regula ferrea AB. . . ." Ibid.; entire description through p. 84, col. 2.

44. Cf. material from slightly earlier in the century in Dear, "Jesuit Mathematical Science."

time."[45] The second "experience," however, is more interesting. Where the first has left unquestioned the truth of the universal experience that the instructions purportedly enable the reader to realize—acting in effect as an elaborate statement of the "proposition" itself—this draws in discrete events to bolster its persuasive effect.

It commences, much like the first, with instructions on establishing the apparatus: "arrange in the plane of the meridian two threads suspended from above from the same point . . . ," proceeding in this way to describe the construction of a sighting instrument. The idea is to have a pendulum the swings of which are counted to mark the interval between the transits of particular stars across the meridian. The number of swings for the same transit interval on different nights are compared, "for if all things have been carried out exactly, and the pendulum moves unvaryingly through its plane, you will find either no disparity in the number of oscillations, or else a contemptible one." Riccioli now supports this claim, apparently unable in his judgment to stand by itself, with an historical narration:

> Thus when we wanted to consider whether the axis of the usual pendulum used for the measuring of astronomical times had suffered any sensible difference by attrition, as of holes E, I [holes through which is slotted the bar that supports the pendulum], or by something else, we counted for three nights the number of oscillations from the transit of Spica to the transit of Arcturus through the same meridian, with witnesses and judges Fathers Francisco Maria Grimaldi and Francisco Zeno, who are most practiced in this matter, actually on May 19 and 28 and June 2, and twice we found the number simply [*simplices*] 3212 and once 3214, having practiced our custom of putting single counters into a dish after each group of thirty oscillations.[46]

The remainder of the chapter largely abandons this momentary citing of participants and witnesses to specific trials. Henceforth all but one of the

45. "Fac enim tibi horologium ex pulvere stanneo breve, putà unius semiquadrantis horae, aut 10 minutorum, ut minus sit taedium in principio experimentorum . . ."; "Si tamen haud ita diligens fueris, unius aut alterius vibrationis differentia esse poterit, inter primi & secundi temporis vibrationes." Riccioli, *Almagestum novum*, Pt. 1, 85, col. 1.
46. "Secundò colloca in plano Meridiani duo fila ex eodem puncto supernè suspensa . . ."; "nam si exactè peracta sunt omnia, & perpendiculum per idem sui planum incessit, aut nullam in numero vibrationum disparitatem deprehendes, aut contemptibilem. Sic nos cùm perpendiculum, ad mensuranda tempora Astronomica, usitatum expendere vellemus num attritione axis, & foraminum E, I, aut aliunde aliquam sensibilem diversitatem passum esset, testibus & adiutoribus PP. *Francisco Maria Grimaldo* & *Francisco Zeno*, in hac re exercitatissimis, numeravimus tribus noctibus, nempe Maij 19. & 28. & Iunij 2. vibrationes à transitu Spiçe ad transitum Arcturi per eundem Meridianum, & bis deprehendimus vibrationes simplices 3212. & semel 3214. usi more nostro singulis calculis, in vas post tricennas quasque vibrationes compositas iniectis." Ibid.

"propositions" appear without comment or with only very brief corrobora-tion, such as Proposition VIII, concerning the relation of pendulum length to frequency (length inversely proportional to the square of the frequency): "Thus by many experiences from Galileo, Baliani and us."[47]

Riccioli's basic technique for establishing the "propositions" he presents in this chapter thus amounts to little more than assertion of his own reliable experience of their truth. The only exceptions involve actual measurement of the period of a pendulum, and these are matters that do not form the subject of a "proposition"—the account of the transit measurements occurs in a proposition concerning isochrony, not actual numbers. The establish-ment of isochrony in the face of possible frictional impediment, together with the uncontested period/length relationship independently corrobo-rated by Galileo and Baliani, were essential in rendering the pendulum suitable as a time-measuring instrument. As such, Riccioli's historical ac-counts describe an exercise in calibration, as an illustration of procedure, rather than discrete experiments, the findings of which were to contribute to an understanding of nature.

The exercise is partially turned to account in chapter XXI, where the accurate measurement of time is essential for overturning what Riccioli argues is a false but increasingly prevalent belief: that all heavy bodies accelerate at the same rate regardless of their weights. This chapter is headed "On the speed of heavy bodies descending with natural motion, and the proportion of increase of their speed."[48] Riccioli claims to have tried to discover the proportion, using pendulums, before the discussions of Gali-leo (in the *Dialogue*) and Baliani (in his *De motu naturali gravium*) ever appeared. He promises to recount his own findings together with his agreements and disagreements with the other two writers.

Proposition I directly challenges the Galilean view: "Of two bodies of equal size [*molis*], that which is heavier descends faster with natural motion from the same terminus to the same terminus, than that which is lighter." He substantiates the assertion with an account of dropping twelve balls, each weighing twenty ounces, made from solid clay, and twelve others made from compressed paper, in pairs from the Asinelli tower in Bologna. The heavier would land first, hitting the ground three pendulum swings before the other, "which," he stresses, "was confirmed by experience re-

47. "Ita ex Galilaeo, Baliano nostrisque pluribus experimentis." Ibid., 86, col. 1. The exception is Proposition XI, again involving exact measurement of the period of a pendulum, and using a historical account of work done with Grimaldi.
48. De Velocitate Gravium Naturali motu descendentium, & Proportione incrementi velocitatis eorum." Ibid., 89, col. 1.

peated twelve times." He is aware that this result contradicts the widely known claims of Galileo and Baliani; the reason for the discrepancy, he asserts, is that they examined the matter using altitudes of perhaps only 50 or 100 feet, whereas the Asinelli tower is 312 feet high. Furthermore, he continues, "Witnesses and judges to this and the following experiences included, although not always everyone simultaneously, Fathers Stephanus Ghisonus, Camillus Roderigus, Iacobus Maria Pallavicinus, Vincentius Maria Grimaldus, Franciscus Maria Grimaldus, Franciscus Zenus, Georgius Cassianus and others."[49]

Proposition II asserts the same thing for two bodies of the same composition but different weight; again an historical presentation appears: "For we dropped, August 4, 1645, from the Asinelli tower, many chalk balls . . . and we always observed the heavier one to strike the pavement" about four feet ahead of the lighter. Riccioli then presents the differences between rates of fall of balls of differing composition (wax, chalk, wood, lead, iron). Proposition III observes that these findings show that Aristotle was nonetheless wrong in asserting (if this is what he really meant) that rate of fall is directly proportional to weight.[50]

Proposition IV maintains the odd-number rule for free-fall, credits Galileo and Baliani with its prior statement, and observes that it has been "often confirmed by our experiences." This locution, reminiscent of Arriaga and Galileo in their stressing of the frequency of trials, receives further accreditation, again reminiscent of Galileo in the *Dialogue*; instead of Salviati and the "Academician" (a fictional collaboration, but apparently no less rhetorically appropriate for that), we have Riccioli and F. M. Grimaldi. They dropped "many chalk balls" from "diverse towers or buildings" chiefly in Bologna—Riccioli names five of them together with their altitudes. They used two very short pendulums for accurate measurement of time, a function accredited by the previous chapter. Now Riccioli wishes to be more precise. His means of accomplishing this is telling: "However, among many experiences, I set forth in the following table two most select ones, the most certain of all, written below."[51] But even these are not to be

49. "Corporum duorum aequalis molis illud quod gravius est, naturali motu citiùs descendit ex eodem termino ad eundem terminum, quam illud quod est levius"; " . . . quod repetito duodecies experimento confirmatum est"; "His & sequentibus experimentis testes, & adiutores interfuere, licet non semper omnes simul, *PP. Stephanus Ghisonus . . .*" etc.: Ibid.

50. "Demisimus enim Anno 1645. Augusti 4. ex Turri Asinellorum plurimos cretaceos globos . . . & semper observavimus graviorem percutere pavimentum. . . ." Ibid.; Ibid., 89, col. 2.

51. "Inter multa autem Experimenta, duo infrascripta selectissima, & omnium certissima . . . propono in tabella sequenti." Ibid., 90, col. 2 (above table).

read as outcomes of individual trials, as Riccioli's exposition makes clear: "Thus in the first experience, when we had observed the aforementioned ball by careful procedure [*operatio*] many times repeated travel from a height of ten feet to the pavement in only five vibrations of the aforementioned pendulum. . . ."[52]

Riccioli's use of tables epitomizes his use of experience in establishing these mathematically dressed "propositions." The first of the tables in these two chapters encapsulates the length/frequency relationship asserted by chapter XX's "Proposition VIII."[53] It coordinates the two variables in columns in a form that might best be characterized as a "ready-reckoner," akin to that other seventeenth-century innovation, logarithm tables, and especially to the much older model of astronomical tables representing the motions of celestial bodies. The latter, represented most notably by the thirteenth-century Alfonsine Tables, the sixteenth-century Prutenic Tables, and the seventeenth-century Rudolphine Tables, were not, it should be remembered, tables of astronomical *data*. Instead, they were generated by geometrical *models* of celestial motions. Similar layouts appear occasionally in texts of the other major classical mathematical science of nature, geometrical optics, as with Ptolemy's and Witelo's tables of refraction. It is now well known that those tables were generated using a simple algorithm to interpolate values between a scattering of empirically determined calibration points; they were not tables recording raw measurements.[54] Riccioli's pendulum table is of just the same kind: the listed values coordinate with empirical determinations through the choice of "typical" touchstone measurements, but their rationale is the exposition of the length/frequency relationship, which they are calculated to exemplify. Similar tables appear on succeeding pages listing number of oscillations of pendulums of particular lengths against astronomical time. The ideal nature of the tables becomes even clearer in light of the presentation, immediately following, of a series of "Problemata" (the standard term for geometrical constructional problems), the solutions to which rely on application of the relationship displayed in the tables.[55]

The tables in chapter XXI have the same characteristics. The information

52. "In primo itaque Experimento, cùm ex altitudine pedum 10. globum praedictum iterata saepius operatione observassemus ad pavimentum pervenire tantum vibrationibus quinque praedicti perpendiculi. . . ." Ibid., 90, col. 1 (below table).
53. Ibid., 86, col. 1.
54. On Ptolemy, Albert Lejeune, "Recherches sur la catoptrique grecque d'après les sources antiques et médiévales," Académie Royale de Belgique *Mémoires,* Classe des lettres, 2d série, vol.52 (1957–58), fasc.2, esp. 152–166; on Witelo, A. C. Crombie, *Robert Grosseteste and the Origins of Experimental Science 1100–1700* (Oxford: Clarendon Press, 1953), 219–225.
55. Riccioli, *Almagestum novum,* Pt. 1, 87, 88, col. 1.

on differing rates of fall of balls of different composition appears in the form of a tabular comparison between each of the pairs, noting which outstrips which and by how far on a standard drop. The figures are straightforwardly empirical insofar as there is no underlying relationship to be established; nonetheless, each number appears as an absolute value rather than as the result of a specific trial, or even as the determined average of a specified number of trials.[56] The table justifying the odd-number rule is itself exactly what one would expect from the pendulum examples: it shows the altitude of drop corresponding to the number of oscillations of a given pendulum, with the final column reducing the figures—for drops up to 240 feet—to display their perfect adherence to the Galilean progression of distance against time. It is followed by a calculation of how long it would take one of these chalk balls to fall from the moon to the earth (a problem borrowed from Galileo's *Dialogue*).[57] The table requires no concept of experimental error, because it does not tabulate experiments. It tabulates the lessons of "experience," and does so according to a well-understood form.

Riccioli's lengthier discussion of the material on falling bodies in Part 2 of the *Almagestum novum* has the same general characteristics. The chapter commences with the undermining of Cabeo, Arriaga, and others discussed above; it continues with sections each presenting a "group of experiences" (*classis experimentorum*) purportedly establishing a particular conclusion. These conclusions, it should be recalled, Riccioli has advertised as being "certain," forming part of "infallible physico-mathematical science."[58] The movement between recipe-like or instructional accounts of contrived experiences, necessarily universal in form although springing from Riccioli's "expertise," and historical accounts of individual trials or sets of trials, again occurs with the implied epistemological priorities just examined, although the greater prolixity serves more powerfully to establish that Riccioli knows whereof he speaks. Even some of the nonhistorical universal "experiences" that Riccioli presents are clearly marked as warrants of the truth of the assertions to which they are attached, rather than as the available grounds for making the assertions in the first place. At the end of three "experimenta" of this kind, for instance, Riccioli writes, "And it is evident by other innumerable experiences of this kind, that a heavy body falling naturally from a higher place has acquired ever greater impetus at the end of [its] motion."[59]

56. Ibid., 89, col. 2.
57. Ibid., 90, cols. 1–2; 91, col. 1.
58. See end of the previous section, above.
59. "Et alijs innumerabilibus experimentis huiusmodi evidens sit, maiorem semper ac maiorem impetum in fine motûs acquisivisse corpus grave naturaliter cadens ex altiore loco." Riccioli, *Almagestum novum*, Pt. 2, 384, col. 2.

As before, the historical narrations are themselves set in a context of justifying more general experiential claims. The chief example, wherein Riccioli and numerous Jesuit colleagues dropped balls from various towers in Bologna,[60] he introduces by saying, "The fifth experience, taken up by us very frequently, has been the distance that any heavy body traverses in equal times in natural fall."[61] The succeeding account includes a diagram of the Asinelli tower where most of the trials were conducted,[62] but the reader is never allowed to forget that these details should not be taken to restrict Riccioli's warrant for his claims. Specifying investigations that he and his collaborators undertook in May of 1640, he adds "and then at other times"—Riccioli's expertise can be seen in his historical narrations, but it is not dependent on them. Riccioli is not adducing evidence for his claims; he is presenting tokens of his experiential knowledge.[63]

Throughout the sections of the chapter, the reader is bombarded with assurances of the frequency with which the experiences were made ("frequently repeated"; "From all these and many other experiences it is evident to us . . ."; and so on).[64] The tables of data do not purport to be complete and foundational; the description of a table of results for bodies falling through water, for example, is described as having been compiled "ex selectis experimentis."[65] Finally, following the individual propositions and their "experimenta," appear "Theorems selected from the foregoing experiences." These are the items of natural knowledge that Riccioli wishes to have accepted on the strength of his expert testimony, such as his argument for the reality of absolute levity. Derivation of these scientific conclusions,

60. See above for Riccioli's earlier presentation of this material; also Koyré, "Experiment in Measurement."

61. "*Quintum* igitur Experimentum sumptum à nobis saepissimè, fuit spatij dimensio, quod grave quodpiam, aequalibus temporibus naturali descensu conficit." Riccioli, *Almagestum novum*, Pt. 2, 385, col. 1.

62. Ibid., 385, col. 2.

63. It is noteworthy that the detailed historical reports of the fourth section, wherein a number of specific dates and named witnesses are presented, and results, sometimes of single rather than multiple trials, are given in a table of actual data for the different descents of paired balls of different weights or compositions, are only adduced "for the unequal descent from the same height through the air of two heavy bodies of diverse weights" ("pro Duorum Gravium diversi ponderis Descensu Inaequali ex eadem altitudine per Aerem": p. 387, col. 1), with some "corollaries" drawn from the data on p. 388, col. 1, to p. 389, col. 1, noting that heavier bodies fall faster except when there is a big counteracting difference in density. There are no correlating measurements of the specific weights of the balls; the actual figures themselves, therefore, rather than the overall trends, have no special significance—they serve only as marks of the sort of thing one finds.

64. ". . . observationem . . . saepiusque repetitam"; "Ex omnibus his & alijs plurimis experimentis nobis evidens est. . . ." Ibid., 390, col. 1; 391, col. 1.

65. Ibid., 390, col. 2.

the distilled "certainties" that he has promised, apparently qualify as "theorems" because they are justified by reference to the previously established experiences (which make the conclusions "manifest," "agree" with them, "show" them, and so forth) much as geometrical theorems are deduced from uncontested premises.[66] Last of all, to complete the mathematical packaging of this piece of physico-mathematical science, Riccioli gives "Problems selected from the foregoing theorems," showing how to put the conclusions to work in such things as comparing rates of fall of bodies in air and in water.[67]

Riccioli, then, uses experience to establish expertise, a quality that he as the author of a text can then employ to undermine the knowledge-claims of others such as Galileo, Arriaga, Cabeo, and Baliani (he repeats his contradictions of them as he lays out his concluding "theorems"). The witnesses he sometimes cites serve to increase the credibility of the experiences described, and thus the authenticity of his own competence to speak. Citation of witnesses would have been irrelevant if Riccioli had expected his readers to create these experiences for themselves. The citations enhance the persuasive efficacy of the text, not the credibility of nature.[68]

Conclusion: Rhetoric, Literary Form, and Epistemology

According to Steven Shapin and Simon Schaffer, the central element of the natural philosophical enterprise of Robert Boyle and the early Royal Society was the category of the "matter of fact." "Matters of fact" were supposed to generate consensus amidst speculative, potentially divisive discord. According to this ideology, everyone could agree on what happens in nature, that is, on the appearances, even if there might be disagreement over causal explanations of those appearances. "Experimental philosophy" therefore meant "natural philosophy of appearances," and those appearances were supposed to be creatable to order through experimental means, and hence cognitively accessible to all.[69] That characterization requires a

66. Ibid., 394, col. 1 to 396, col. 1.
67. Ibid., 396, col. 1 to 397, col. 2.
68. Cantor, "The Rhetoric of Experiment," 169–173, discusses the role of nature as textual interlocutor in Galileo.
69. Issue is taken with this view by Sargent, "Scientific Experiment and Legal Expertise." However, actual (textual and social) experimental practice, rather than programmatic and methodological statements by Boyle, Sargent's focus, bear out the characterization presented here; cf. Shapin and Schaffer, *Leviathan,* chap. 2.

further refinement, however: an experiment—or more properly, an experimental report—need not necessarily be able to stand for "what happens" under designated circumstances. Logically, it might function only to stand for "what happened" on a *particular* occasion. For the singular experiment to stand for the universal experience, an appropriate kind of argumentative framework needs to be in place, explicitly or implicitly, within which it can play that role.

Boyle did not use axiomatic deductive argumentative structures, which were supposed to constitute true science in an Aristotelian sense, whereas Galileo and the Jesuits did. The difference is of great significance. Boyle did not employ an Aristotelian notion of experience; they did. Boyle reported singular historical events; they needed universal statements of behavior even when giving historical accounts by way of collateral. The deductive, demonstrative model of natural knowledge meant that empirical statements had to play the part of axioms; that is, they had to look like universal statements of the way everyone knows things are—like geometrical axioms. The probabilistic model of Boyle, on the other hand, required a category of the "matter of fact," the legitimacy of which depended precisely on accredited, and therefore specifiable, occurrences.

When Riccioli (or, to reinforce his argument in the wake of criticism, Arriaga) cited specific times and places for events related to the establishment of a philosophical claim, the textual effect that he sought differed radically from the one sought by Boyle. Boyle wanted to convince others that something had happened; Riccioli wanted to convince them that he, the author, knew what he was talking about. The one aimed at vindicating a historical assertion, the other at vindicating a personal, and authoritative, quality. Making knowledge of nature from true histories was, of course, considerably more difficult than making it from a position of expertise, of familiarity with nature. Riccioli could take for granted a set of epistemological assumptions shared by his readers that gave his discourse its meaning; Boyle and his allies lacked such a framework, which is why they so frequently characterized their work as a Baconian collecting of facts—there was no clear way forward to making universal knowledge about the structure of nature. Even the simple inductive move from the accreditation of a single event to a generalization about how such situations *always* behave was strictly unwarranted without the means to invoke an epistemologically justifiable methodological model.[70] In that sense, Fabri's solution was the

70. There is an important difference here between the establishment in a legal context that an event occurred, and the same establishment in a philosophical context. A law trial validates

most straightforward of all: he denied the foundational role of experiment—that is, recondite and particular, because precise, experience—altogether.[71]

To understand the appearance of experiment, or experimental science, in the seventeenth century, then, we have to look at the general category of experience as an element in formulating natural knowledge. That means investigating the literary constitution and function of experience in scientific argument, because it is in texts that the knowledge is made. Episodes such as Galileo's rolling of balls down inclined planes, or Riccioli's raining of clay balls from Bolognese towers, do not in themselves amount to experiments. They have meaning only by virtue of an entire set of cognitive assumptions and expectations appropriate to their intended readership. In the first half of the seventeenth century, that framework can best be described, at least as far as the role of experience is concerned, as Aristotelian. The creation of experimental natural philosophy must be understood through the exploitation and deformation of that framework by literary techniques and strategies.

The completion of this work was supported by National Science Foundation grant DIR-8821169.

an assertion that, on a particular occasion and involving particular people, a particular event occurred. The kind of event, however—murder, or theft—is already well known; the question of whether such things happen at all is not at stake. But an experimental event is by its very nature *novel*; its establishment is a matter of determining the reality of a kind of event that is not already generally accepted as occurring. This caveat should be borne in mind in reading Sargent, "Scientific Experiment and Legal Expertise." Newton's views on how to create universal scientific statements from singular experiments were a new departure: see Zev Bechler, "Newton's 1672 Optical Controversies: A Study in the Grammar of Scientific Dissent," in Yehuda Elkana, ed., *The Interaction Between Science and Philosophy* (Atlantic Highlands, N.J.: Humanities Press, 1974), 115–142; Paul K. Feyerabend, "Classical Empiricism," in Robert E. Butts and John W. Davis, eds., *The Methodological Heritage of Newton* (Toronto: University of Toronto Press, 1970), 150–170. The related difficulty of establishing the "sameness" of diverse experimental events reported from different sources (as with reports sent to the early Royal Society), and Newton's behavior in that regard, is discussed in Steven Shapin, "'The Mind in Its Own Place': Science and Solitude in Seventeenth-Century England," unpublished MS, section "The Science of Solitude," subsection (b).

71. This attitude should be compared with that of Hobbes, as discussed in Shapin and Schaffer, *Leviathan*, chap. 4.

Frederic L. Holmes

6. Argument and Narrative in Scientific Writing

IN A RECENT History of Science Society Lecture[1] I outlined the view that a scientific research paper necessarily combines elements of narrative and elements of argument. This duality of literary form reflects the dual function of the research paper as both a summary of the author's current *findings* and an account of investigative work that she or he has carried out during a previous period of time. One or the other of those two elements may predominate in the organization of the paper, but the other element is, I suggested, bound to be woven into its structure, however inconspicuous or indirect its appearance may be. Here I would like to develop this position somewhat further and to illustrate it by discussing a cluster of papers, written around 1700, that I believe to be representative of a formative stage in the emergence of the research paper as a stable genre in the scientific literature.

There is a common belief—of which a vivid expression appeared recently in an article in the *New York Times* Review Section entitled "Scientific Writing: Too Good to be True?"—that modern scientific writing is quite unlike that of the past. Pressed into a "conventional format," the scientific article of today is deprived of all subjective elements, stripped of all that is extraneous to the conclusions reached, and shorn of the human activity underlying the conclusions it presents. This style, which the author of the *New York Times* article likened to "literary Novocaine," is contrasted with the lively spontaneity supposed to characterize early scientific writing, writing that could be exciting, that "included something about the nature of scientific endeavour—its difficulties, the prospects for failure and the flexibility necessary to do scientific work."[2] In short, the modern scientific paper is perceived as coldly analytical, scientific writing of the past as free to tell adventurous "stories" about exploring nature.

1. Frederic L. Holmes, "Scientific Writing and Scientific Discovery," *Isis* 78 (1987): 220–235.
2. Bob Coleman, "Science Writing: Too Good to Be True?" *New York Times* Book Review, September 27, 1987, p. 7.

Although it is obvious that some of the range of expression allowable in the scientific writing of the past finds no place in the specialized scientific literature of today, there is, I will argue, a strong continuity of both form and content linking what we call today the "research paper" with practices extending through the past three centuries. The "conventional form" imposed on the modern journal article is not a departure from, but the outcome of the long evolution of a form that emerged during the late seventeenth and early eighteenth centuries along with the learned journal as the forum for reporting scientific investigations.

Recently Steven Shapin and Peter Dear have provided two penetrating interpretations of the origin of what Dear refers to as the literary form of the research report in seventeenth-century England. Both focus on Robert Boyle, while Dear extends his analysis to practices characteristic of the members of the early Royal Society. Both argue that Boyle and other "virtuosi" resorted to highly detailed narrative accounts of their experiments and observations as a means to attain *authority* for the assertions they based upon them. Shapin emphasizes that the provision of dense "circumstantial details" substituted for the actual witnessing of the experiments that would ideally be required (and was sometimes done) in order to assure the relevant community that the experiments had been performed as claimed. Boyle's well-known verbosity in describing his experiments was, according to Shapin, in part a deliberate plan "to give the impression of verisimilitude," and thereby to "compel assent" to "matters of fact."[3] Dear stresses that Boyle and other Fellows of the Royal Society filled their descriptions of experiments and observations with contingent details of time and place in order to convey the impression that these were actual events, of which the authors were the faithful reporters. "The described experience was *discrete*: it was a single, historical occurrence, not a generalized statement." In Dear's view this manner of presentation was intended to transfer to direct observational experience the authority that had formerly been lodged in traditional texts.[4]

Boyle attempted to maintain a sharp distinction between reporting and interpreting. In the prefatory letter to his *New Experiments Physico-Mechanical touching the Spring of the Air*, he called his descriptions of the experiments "narratives," and wrote that he had "left a conspicuous interval" between them and the "discourses made upon the experiments," in

---

3. Steven Shapin, "Pump and Circumstance: Robert Boyle's Literary Technology," *Social Studies of Science* 14 (1984): 487–494.

4. Peter Dear, *"Totius in verba:* Rhetoric and Authority in the Early Royal Society," *Isis* 76 (1985): 145–161.

order that readers "who desire only the historical part of the account" need not read "the reflections too."[5] He viewed his treatise, therefore, as clearly divisible into historical narratives and discourses, or reflections. In practice these divisions were not always clean. For example, in the famous chapter "Two new experiments touching the measure of the force of the air compressed and dilated," Boyle recounted the experiments not only in picturesque detail, but in a lively narrative that included such subjective asides as "we observed, not without delight and satisfaction, that the quicksilver in that larger part of the tube was 29 inches higher than the other." Elements of discourse based on the results crept easily into the "historical part" itself, however. The sentence immediately following this one stated, "Now that this observation does very well agree with and confirm our hypothesis, will be easily discerned by him that takes notice that we teach, and Monsieur Pascal and our English friends' experiments prove, that the greater the weight is that leans upon the air, the more forcible is its endeavour of dilatation." After developing this argument, which incorporates evidence supplementary to that of the experiment described, and which may or may not have been on Boyle's mind at the point at which he made the observation that so satisfied and delighted him, he slipped back into his narrative: "We were hindered from prosecuting the tryal at that time by the casual breaking of the tube."[6] Not only are elements of discourse thus interspersed with those of narrative, but the chapter as a whole is situated within Boyle's extended argument against the "opposing" hypothesis of his "adversary," Franciscus Linus. The experiments that we treat as the foundation of Boyle's law are invoked in support of Boyle's position that Linus's funiculus hypothesis was not only unintelligible, but unnecessary to explain the known phenomena.[7] Despite his own intentions, narrative and argument are inextricably mingled in his account of these landmark experiments.

Central as Boyle and the Royal Society may have been in setting out the criteria for reporting experiments and observations in a form that made them acceptable as "matters of fact," the institution which I believe contributed most strategically to the creation of the scientific research paper was the French Académie des Sciences. The papers read by members of the Academy to their fellow Academicians at their weekly meetings and published subsequently as the *Mémoires* of the Academy became, by the early eighteenth century, the most consistent exemplars of the form that has

5. Robert Boyle, *New Experiments Physico-Mechanical touching the Spring of the Air, and its Effects* (Oxford: H. Hall, 1662), [A4].

6. Ibid., Pt. 2, 58–59.

7. Ibid., Pt. 2, 57–68.

persisted as the modern research paper. This development was the result not only of the nature of the forum and the format in which Academicians presented their results, but of the fact that the very nature of research as sustained experimental investigation to be reported in such fashion was largely a product of the organization of the Academy.

The features of the Academy of Sciences most pertinent to its role in these developments are highlighted in Roger Hahn's *The Anatomy of a Scientific Institution*. The Academy adopted from the beginning a policy of including within its ranks only those who were willing to become scientific professionals. The salaries that the Academicians received allowed them "the luxury of being full-time researchers," a luxury then accorded to very few anywhere else. Their positions not only permitted but virtually required the various Academicians to become specialists in the branches of science represented among them. Hahn points out that the Academy maintained the ethos that new and secure knowledge could not be attained quickly; it would emerge only gradually, through the slow accumulation of "unassailable facts."[8] It was their professional security that allowed the Academicians not only to profess such a viewpoint, but to act upon it by undertaking investigative projects of long duration—investigations requiring a daily effort lasting not for weeks or months, but for years or decades.

As is generally known, during its first decades the Academy organized its members into several collective investigative projects, supposing that they could in this way attain more certain knowledge, freed from individual human biases and misjudgments.[9] If we view their approach from the standpoint of Shapin's and Dear's treatment of the approach of the English virtuosi, we can see these projects as an alternative means to substitute a new form of authority for the rejected forms of traditional authority. Instead of attempting individually to gain community assent by the fullness of their narrative accounts of observations and experiments, the Academicians hoped to achieve concensus by exercising collective, in place of individual, judgment. Among the projects carried on for nearly thirty years was a monumental "History of plants," which was claimed to be a "work of the whole Academy." The project incorporated several subprojects, including the collection and external descriptions of plants, examination of their medicinal properties, and the chemical analysis of their matter.[10] I would like to focus on this latter aspect of their work.

8. Roger Hahn, *The Anatomy of a Scientific Institution: The Paris Academy of Sciences, 1666–1803* (Berkeley: University of California Press, 1971), 32–33, 51–52.

9. Ibid., 24–26.

10. Denis Dodart, "Mémoires pour servir à l'histoire des plantes," *Mémoires de l'Académie Royale des Sciences* 1666–1699, vol. 4 (Paris: Imprimérie Royale, 1731), 121.

One of the chemists among the original Academicians, Claude Bour-delin, "executed and conducted almost all of the chemical operations . . . and maintained most of the registers."[11] These bound notebooks have survived. They contain detailed records of each of the laborious distillation experiments performed on every kind of plant material the Academy could acquire.[12] They constitute the earliest example with which I am familiar of the systematic maintenance of experimental records in the form that has since become the standard type of repository for laboratory investigations. Such a record was, in fact, necessitated by the very scale of the project on which the Academy embarked.

The project began in 1668. A few preliminary reports of its progress appeared in the summaries of meetings in subsequent years.[13] In 1676 Denis Dodart wrote a comprehensive memoir describing the work so far achieved, claiming to have incorporated into his discussion the views and reflections of all of the Academicians who had contributed to the project. "We have decided," he explained, "to make our project public, to give an account of the success of the experiments that we have done, and to propose what we believe should be done in the future."[14] The account began with a narrative statement: "At the time the Academy undertook to write the natural history of plants, it did not ignore the extent and difficulty of its plan."[15] Dodart did not continue in this mode, however. Instead, he organized his discussion topically, providing a remarkably penetrating analysis of the nature of the problems involved and the general modes of attack on these problems that the Academicians had considered and imple-mented. The largest section of the overall memoir dealt with the chemical analysis of plant matter. It began with a summary of the various points of view that had been expressed by previous chemists and the inadequacies left by earlier analyses. Dodart presented an extensive "reflection" on the use of fire in the analysis of plants, in the form of general principles that ought to guide such an investigation from the start, even though these consider-ations had actually grown out of the experience the Academicians had gained over the course of their investigation. Similarly, Dodart discussed in

11. Idem.

12. Claude Bourdelin, *Recueil d'analyse chimique,* Cahier Bourdelin, Fr. Nouv. Acq. 5133–5149, Bibliothèque Nationale, Paris.

13. *Histoire de l'Académie Royale des Sciences* (Paris, 1733), vol. I, 18–20, 56–58, 120–122, 161–162. See Frederic L. Holmes, "Analysis by Fire and Solvent Extractions: The Metamorphosis of a Tradition," *Isis* 62 (1970): 133–136.

14. Dodart, "Histoire des plantes," 124.

15. Ibid., 123.

logical rather than chronological sequence the various methods that the Academicians had learned to employ, over these years, to identify the products of their distillation analyses. This discussion of methods occupied over four-fifths of the eighty-four pages of the portion of the memoir dealing with the chemical analyses. In the last few pages Dodart gave "some idea of the consequences one can draw from all of these researches";[16] that is, in tentative form he outlined some of the general differences the Academicians had found in the results obtained from the analyses of plants of differing properties. One could easily subdivide Dodart's memoir according to the headings—"introduction," "methods," "results," and "conclusions,"—so familiar in the scientific papers of more recent times.

Although the dominating structure of Dodart's memoir was that of a critical argument rather than a narrative, there were nevertheless scattered through it numerous references to events that had occurred during the course of the Academy's long investigation. Here are a few examples:

We have found that certain spirits that we call mixed . . . reddened a solution of German vitriol.[17]

It is true that we found some very acrid spirits which reddened tournesol, but it appears that it is not on account of their acidity that they redden.[18]

We wanted to imitate the nature of the [mixed spirit] . . . by mixing acid and sulfur [alkali] in different proportions. But the mixture always exerted the effects either of an acid or of a sulfur. . . . That led us to suspect that the acid and the sulfur are mixed in this liquor not only in a [certain] proportion, but in a peculiar manner.[19]

There were also many indications that Dodart's memoir constituted a progress report on investigative work that remained incomplete and that was expected to continue. For example:

We have not yet worked enough on the oils to give the details of their composition here. All that we can now say is that . . .[20]

All of these experiments are not yet in a state in which we can consider that more than a beginning has been made.[21]

16. Ibid., 225.
17. Ibid., 193.
18. Ibid., 195.
19. Ibid., 219.
20. Ibid., 219.
21. Ibid., 223.

In this memoir emanating from the French Academy we see a very different balance between argument and narrative than in Robert Boyle's chapter on the measure of the force of the air. Detailed narratives of experiments dominate Boyle's treatment, subordinating the arguments the experiments serve. In Dodart's memoir, nonnarrative analyses of the nature and methods of the investigation so preempt the organization that no coherent descriptions of individual experiments appear. Narrative details intrude repeatedly in the midst of these topical discussions, but they are not organized into a story. The contrasts between these two exemplars of seventeenth-century scientific literature reflect more than differences of style. They are adaptations to the very different nature of the experimental work described and the circumstances under which the work was performed. Boyle based his conclusions on two experiments, both of which had been difficult to carry out, but which were compact enough to describe fully as discrete events. It would be impossible to provide such a description of the myriad discrete events composing an investigation that had already lasted for several years. In a revealing passage Dodart explained why he had not described individual experiments:

> We record in the registers of the analyses, all of these substances with their differences. . . . We record these analyses as a type of *proces verbal;* we note how many times one has changed the receiver, we describe the products of the distillation in detail, that is, the weights and sensible properties of these products; we note the time that it has taken to distil each of these products and the degree of fire, . . . because we believe that someone will be able to elicit some new knowledge from these particulars, or find the occasion for some new research, and because it is not possible to keep these registers in any other way, when one wishes to record matters at the time they take place. But we also believe that we ought to relate all of these particulars to certain main principles which assist the memory and save the mind from the confusion into which this great multitude of circumstances would plunge it.[22]

This passage identifies one of the factors that very early forced a departure from the dense narrative accounts of experiments and observations that Dear and Shapin have shown to have been so important in the Royal Society. When experimentation could become a full-time occupation, and trains of individual experiments began to coalesce into prolonged research, the accumulation of data could easily overwhelm the possibility of describing it fully in the order in which it had been gathered. To master the results

22. Ibid., 185.

of such investigations it became necessary to reorganize the investigations in a synoptic form that focused on general principles and procedures rather than on the contingent circumstances of time and place that marked each experiment as a real event.

A second factor that may have underlain the differences in the form of memoirs from the French Academy and the writings of the virtuosi is that the Academicians did not feel the need that the latter had, according to Shapin and Dear, to employ dense narrative description in order to acquire authority for their conclusions. Although the Academicians did eventually make their work available to a broader public, they addressed themselves in the first instance to one another. They were often able, in fact, to witness directly the operations conducted by their chemist. Those who had not been present for these could consult the laboratory notebooks if necessary. Whether the experiments had actually been performed as claimed was therefore not a serious issue for the Academicians, and they could concentrate their attention on broader questions of interpretation.

Dodart's memoir turned out to be the high point in the task of plant analysis. The project continued for two more decades, until Bourdelin died, but with the passing years it became increasingly routine. Here, too, the Academy set a precedent, encountering a problem to which long-term research carried out by full-time professionals has ever since been vulnerable. By the time the project ended, the Academy had already abandoned the principle of communal inquiry from which it had originated, in favor of individual investigative activity. The new regulations of 1699 codified this change, requiring each Academician to "choose a particular object of study, and by his accounts to the assembly," to "try to enrich all those composing the Academy through his wisdom, and to profit from their remarks."[23]

By this time also the Academy had appointed a second generation of chemists, who rejuvenated its activity in this field. In the 1690s two veteran chemist-apothecaries, Nicolaus Lemery and Moise Charas, entered its ranks, along with the able younger man Wilhelm Homberg, who had studied with Boyle. A little later these were joined by another pharmacist-chemist, Simon Boulduc, and three ambitious aspiring chemists, the brothers Etienne-François and Claude-Joseph Geoffroy, and Lemery's son Louis. As Robert Multhauf has pointed out, the Academy had now assembled together the leading chemists in France.[24] They fulfilled energetically their

23. Hahn, *Anatomy of a Scientific Institution*, 29–30.

24. Robert P. Multhauf, *The Origins of Chemistry* (New York: Franklin Watts, 1966), 272–273.

responsibility to pursue individual investigations of particular problems within their common field.

The accounts of their research that these chemists presented to their assembled colleagues appeared regularly afterward in the Memoirs of the Academy of Sciences published annually (often with considerable delays) after 1700. Diverse as the memoirs were in detail, they shared certain general characteristics. Most of them were short, ranging between two and ten pages. They therefore treated their subjects compactly, in keeping with their function, for these papers were not comprehensive treatments of broad problem areas (Homberg's *Essays* on the chemical principles were in this regard exceptional), but reports of limited investigations oriented toward special questions. Typically, their authors presented at intervals several papers representing successive stages of continuing lines of investigation. I would like to focus now on a group of such papers read to the Academy between 1700 and 1710. These papers, along with their analogs in other fields represented in the Academy during this period, helped to establish a format that has continued ever since to be characteristic of scientific research papers.

The most prolific investigator of the group during this decade, Wilhelm Homberg, published at least seven papers entitled "Observations" on topics within plant or mineral chemistry. Each of them began with a statement of a particular problem, moved into Homberg's efforts to solve the problem through an experimental investigation, and ended with a statement of conclusions reached, an explanation of the phenomena observed, or a discussion of implications of his results. The patterns in which these elements were woven into a coherent paper varied considerably, however, from paper to paper.

"Observations on the oils of plants," read to the assembly in 1700, dealt pragmatically with a restricted operational problem. The quantity and quality of the oils obtainable from plant matters varied greatly according to the procedures used to extract them. Homberg's paper was essentially a report of the experiments he had carried out in an effort to obtain more of the oil contained in the plants than ordinary distillation methods yielded. He did not describe all of his experiments, however:

> I have made numerous trials in order to discover a convenient method . . . ; I shall report here only those that have succeeded, neglecting those that either did not succeed at all, or that still require further work to be perfected.[25]

25. Wilhelm Homberg, "Observations sur les huiles des plantes," *Mémoires de l'Académie Royale des Sciences,* 1700, 2d ed. (Paris, 1719), 213.

Such statements, which appear repeatedly in the memoirs of the Academy's chemists, refute the stereotyped view that it is only scientists of today who regard it as superfluous to discuss their failures.[26] After summarizing some observations that had led him to think that the acid or the volatile salt contained in the plant matter might aid in the extraction of its oils, Homberg barely mentioned certain experiments that had clarified the situation for him sufficiently to lead him to fix his attention on the action of acids. The next group of experiments he described in narrative style. "The first experiment that I made," he wrote, "was to mix distilled vinegar with the thick fetid oils of several plants." He described such an experiment in a condensed step-by-step form. Deciding that the vinegar contained too little acid to have "a great effect on the oil," he said, "In my second trial I mixed one part of spirit of salt [or marine acid] with two parts" of the oil. Again enumerating the steps in the distillation and describing the products, Homberg related, "I was sufficiently convinced by this second trial, that strong mineral acids can unite with the oily matters of plants without destroying them." Because the smell of the fetid oils on which this process had succeeded rendered them impractical (for medical purposes), however, he said, "I abandoned them and continued my experiments on the essential or aromatic oils." Mixing spirit of salt with fennel, a plant which yielded an essential oil, and allowing them to ferment together before distilling the material, Homberg obtained one third more oil than he did when he carried out the same processes without the acid. He "was persuaded" by these results, he concluded, that spirit of salt not only augmented the fermentation, but rendered the oily parts more liquid, so that they were "more easily removed by the heat."[27]

This relatively simple memoir is in several ways typical of those written by Homberg and his fellow chemists in the Academy. It is both an argument and a report of an investigation, the two aspects corresponding sufficiently so that Homberg could present them both in a predominantly narrative form. It was not, however, a "dense" narrative in the sense that Shapin and Dear have defined those of Boyle and his Royal Society colleagues. Homberg gave no extraneous details concerning his experiments. He mentioned neither the time nor the place in which he had performed them, nor did he describe the apparatus he had used. He supplied just enough information so that readers well acquainted with the standard procedures for analyzing plant material would be able to follow what he had done. He omitted unsuccessful trials and details of experiments he later

26. See Coleman, "Scientific Writing."
27. Homberg, "Huiles des plantes," 212–217.

regarded as preliminary. At the heart of his investigation he appears to have been recapitulating it as he had carried it out; given the simplifications he acknowledged in his description of earlier phases, however, we would be prudent to regard even this central portion as a somewhat idealized version of the actual events. Throughout his treatment he selected those details he deemed most pertinent to the conclusion he reached.

Other memoirs by Homberg display interesting variations on the pattern shown in this one. "Observations on the volatile salts of plants," which he presented in 1701, began with what Homberg termed a "paradox." The distillation of unfermented plants sometimes yielded a reddish liquid that "shows simultaneously the qualities of an alkali and an acid"; yet "the idea that we have concerning acids and alkalis is that they cannot come into contact without destroying one another." Homberg then led his readers through what appears to be the main steps of the trail he had followed to solve this puzzle. "I supposed," he wrote, "that the reddish liquor contained also a portion of an oil which could prevent the acid and alkali from acting on one another. . . . Nevertheless, the fact appeared so extraordinary to me that I was curious to examine it with care. I therefore took two plants, one containing much volatile salt, the other much acid." After performing various operations on the matter of each plant to obtain their respective acid and alkaline spirits, he poured these two liquors together and found that they did not produce the ebullition characteristic of an acid-alkali reaction. "I was very surprised that this mixture of the two liquors produced no movement," he wrote. "That showed me that my first conjecture was not correct." Trying next to separate by distillation the two liquids he had combined, he obtained by sublimation a white crystalline salt. "Knowing no one who had mentioned" such a salt, "this salt appeared to me to be completely new."[28]

Homberg tried the new salt on a person with a kidney pain, but it did not help. He described the properties of the salt—a kind of sal ammoniac in which distilled vinegar replaces the common salt of ordinary sal ammoniac—without detailing the experiments through which he had discovered these properties, probably because they were routine procedures. The resemblance between the volatile salt he had used in producing this new salt and the volatile salt in urine led him to believe he could produce the same salt by combining the latter with distilled vinegar. He therefore repeated the operation in this manner and again obtained the new salt. After describ-

28. Wilhelm Homberg, "Observations sur les sels volatiles des plantes," *Mémoires de l'Académie Royale des Sciences,* 1701, 2d ed. (Paris, 1719), 221–222.

ing further experiments intended to produce the salt in other ways, and to find out why the acid and alkali he had first used had not reacted in the usual way, Homberg ended his paper with an explanation both of the reaction and of the ability of an acid and alkali to exist together. Both explanations were speculations about the shapes of the particles of the two substances. Such mechanistic explanations of observed chemical operations were typical of the conclusions Homberg drew in a number of his papers, and in this case they enabled him to modify his "idea" of acids and alkalis so as to resolve the paradox with which he had begun.[29]

We cannot be certain that Homberg did not also in this case omit parts of his investigation that appeared afterward unessential to his conclusions. Nevertheless, he did cast his paper in the form of a story that appears to reveal the course of his interlocked reasoning and operations from beginning to end. It is probable that he related it in this way because the unanticipated events during the course of the investigation seemed inseparable from its outcome. It was through the unexpected turns that he was led to discover a salt he had not set out to find.

In other papers Homberg presented his work in forms that were partly narrative, partly synoptic. In his important "Observations on the quantity of acids absorbed by alkaline earths," presented in 1700, he described the problem, mentioned previous related experiments he had performed, related how he had prepared two acid spirits to use in the new experiments, and enumerated fifteen kinds of alkaline earth materials that he had dissolved in these acid spirits. Then he simply gave a list of the weight of each material dissolved, respectively, by one ounce of spirit of nitre and one ounce of spirit of salt. These tables thus summarized the outcome of at least twenty-two different experiments. There was presumably no need to describe individual experiments because he had followed the same routine procedures in each of them. The bulk of his paper he devoted to a discussion of the results and their implications.[30]

Homberg organized his paper "Observations on the analyses of plants," in 1701, mainly as a discussion of the long-debated question of whether the products of a distillation analysis had actually existed in the plant matter or were creations of the fire. "I have made numerous experiments to clarify these doubts," he wrote, "but I shall give here only one example to indicate the manner in which I proceeded, and then we shall draw our conclusions."

29. Ibid., 222–225.
30. Wilhelm Homberg, "Observations sur la quantité d'acides absorbés par les alcalis terreux," ibid., 1700, 64–74.

Describing compactly a set of three distillations of grape juice, one using the juice fresh, another using juice that had first been partially evaporated and exposed to gentle heat, the third using fermented juice, Homberg then discussed at length his reasons for believing that the latter two processes were better suited than the first to "discover the true principles and virtues of the plant."[31] In this case Homberg did not describe the course of an investigation. He picked out from an investigation a generalized example of the type of experiments he had pursued, and we have no way of knowing at which stage in a longer sequence of experiments he had reached these results. There was little correspondence between the order of his argument and his sequence of research, so that his paper incorporated only a small fragment from an otherwise untold narrative.

In 1706 Homberg presented a paper entitled "Observations on iron and the burning glass." The first two pages of the paper described in considerable detail what occurs when small particles of iron are exposed to the heat of the sun focused through a magnifying lens. The description must have been based on experiments that Homberg had performed, but he described the process in generic form, as what happens whenever this is done. Then he interjected a brief reference to an investigative event: "Chance led us to discover that in all ashes there is found a blackish powder that is true iron." Instead of relating the particulars of his discovery, however, Homberg gave directions for how "one can verify" it. "Burn whatever dry herbs or wood you wish to ashes. Take [certain] precautions." Follow several further procedures, and "you will obtain a grain of iron," and so forth.[32] By a linguistic device, therefore, Homberg presented his experiments not as singular events but as repeatable findings, reliable enough that anyone who carried out the same operations would observe the same phenomena he had observed. The narrative of an investigation was obscured by the form of presentation, reasserting itself only at that point at which "chance" had intervened in otherwise orderly procedures.

If we were to extend this discussion to the other chemists of the Academy during these years we would find, aside from any stylistic idiosyncrasies, a similarly diverse array of patterns connecting the elements of reporting and of argument in their research papers. I shall mention just one, a memoir presented by Etienne Geoffroy in 1707, called "Clarifications concerning the artificial production of iron and the composition of other

31. Wilhelm Homberg, "Observations sur les analyses des plantes," ibid., 1701, 115–119.
32. Wilhelm Homberg, "Observations sur le fer au verre ardent," ibid., 1706 [1707], 158–165.

metals." This paper was essentially a defense of the proposition that the iron that he and other chemists had obtained from a mixture of linseed oil and clay was "a new production, a compound resulting from the combination of several principles contained in the materials which yield that metal." Geoffroy structured his paper in the form of a series of objections to this view and his responses. The responses incorporated many observations based on chemical operations and experiments, but they did not appear explicitly as experiments that Geoffroy had performed. They were woven into the structure of his argument as items of evidence. At one point Geoffroy brought to bear observations contained in the writings of Boyle, and some of the other observations must also have been either common knowledge or drawn from the work of other chemists; however, much of it must have been the outcome of his own investigation, for he stated at the end, "I have reported here only what my researches have taught me; only time and our experiments [expérience] can instruct us about the rest."[33]

The context within which Geoffroy contributed his paper explains why he cast it entirely in the form of an argument, leaving no vestige of a description of the investigations which his closing remark suggests must have lain behind it. He was participating in a lively debate among the chemists in the Academy, who were divided over the question of whether the iron that Geoffroy had earlier been able to extract from plant matter was formed in the process,[34] as he claimed, or preexisted in the plant. Since the paper was part of an ongoing argument, it was natural for him to cast it in this form.

Each of the papers I have described includes, implicitly or explicitly, elements of argument and elements of narrative; however, the samples I have chosen obviously span a spectrum, from a paper that appears to follow the course of an investigation to one in which the investigation is all but buried in the argument. I have suggested that such variations may not be arbitrary, but adapted either to the character of the investigation reported or to the nature of the argument to which it was addressed. These memoirs, therefore, display a great deal of flexibility within the bounds of what I have asserted to be the format of prototypical "modern" research papers. Both the format and the flexibility reflect the particular circumstances within which these Academicians pursued their work: that is, the organizational

33. E. F. Geoffroy, "Eclaircissements sur la production artificielle du Fer, et sur la composition des autres Métaux," ibid., 1707 [1708], 176–188.

34. I have discussed this debate in an unpublished manuscript dealing with the early history of plant and animal chemistry.

structure and membership of the Academy of Sciences and the broad investigative enterprise that the Academy nurtured.

The chemists of the Academy were freed from the need to describe their experimental investigations with all the detail they could muster, because they formed a community of specialists. Each chemist was communicating primarily with other chemists who understood the basic procedures involved and shared many of the same assumptions. They comprised what the belatedly honored Ludwig Fleck has called a "thought collective."[35] It was both possible and convenient, therefore, to describe experiments in foreshortened, selective form, mentioning just those conditions, sequences of operations, and observations that others with similar experience needed to know in order to grasp the significance of whatever was novel in the work presented. When the author described particular events that had occurred during the course of an investigation, it was because they were unusual or significant to the outcome. The routine could be omitted. The repetitive could be reduced to general descriptions of procedures. Failed or preliminary experiments did not need to be described to enhance the verisimilitude of the description, because the author's colleagues would take it for granted that these were inevitable accompaniments of innovative research.

Within this setting it was not only reasonable, but necessary, to omit from the papers presented whatever in the investigative record did not appear pertinent to the argument for which the author wished to use it. On the one hand, like their predecessors who had worked communally, this second generation of Academic full-time chemists could individually amass more experimental information than they could usefully report. Selectivity and reorganization were demanded. On the other hand, the form of publication on which the Academy had settled by 1700, collections of memoirs appearing annually, restricted the space available for each individual memoir while encouraging sequences of limited progress reports on ongoing investigations. That was just the sort of sustained investigative enterprise that their secure professional positions enabled the Academicians to pursue. The result was the consolidation of a literary genre. Although less strictly codified than the research paper of the twentieth century, the memoirs published by these chemists in the early eighteenth century bear the clear resemblance to them that one would expect of a recent ancestral form.

I have discussed the chemical papers published in the Memoirs of the Academy rather than those in botany, zoology, mechanics, or other subjects

35. Ludwig Fleck, *Genesis and Development of a Scientific Fact* (Chicago: University of Chicago Press, 1979), 38–51.

that might have illustrated the same general points, because I happen to have studied these papers previously for another purpose. They are, however, peculiarly well suited to illustrate my theme. More thoroughly than any other early science, chemistry has from its inception been embedded in experimentation. It was so by definition, since its domain is one of manipulated rather than of natural events. Any scientific research paper has a past history of thought, of gathering either ideas or information. In theoretical papers, however, that which has been discovered or found through a temporal process can more nearly merge with that which is presented and argued. Experimental papers transform into verbal form findings which have been reached through activities involving materials, physical apparatus, and operations carried out at discrete times and places. The experimental paper thus is necessarily in part a report on these events. Sustained experimentation has an intrinsic chronological structure which makes narrative the natural mode in which to describe it. To incorporate what is salient from the investigative "story" into a paper structured as a critical argument requires transformations that are as central to the creativity of science as are the investigations themselves.

The conventional form of the modern scientific paper appears at first sight to suppress the personal narrative of an investigative enterprise along with other manifestations of subjectivity. The standard sections—introduction, methods, results, discussion, and conclusion—seem to be logically ordered components in an analytical structure, not stages in the unfolding of a story. The modern scientific reader, or editor, it may be presumed, has no interest in how the scientist has arrived historically at the conclusion placed at the end of the paper, but in the evidence he or she can now provide in support of that conclusion. We can, however, view these standard sections of the paper also as a framework within which to fit a heavily stylized, formulaic version of a story.

The two faces of this format can be elicited even from contemporary manuals on scientific writing. A well-known book by Edward J. Huth, *How to Write and Publish Papers in the Medical Sciences*, asserts, for example, that scientific papers "are built on the principles of critical argument"—"they argue you into believing what they conclude." A scientific paper should state a problem, present evidence, weigh supporting or conflicting evidence, and reach a conclusion. Such criteria would appear to leave little space for a narrative of the historical investigation on which the paper is assumed to be based. A little further on, however, Huth adds that "Well-written papers use the natural sequence of the research 'story' but with the elements of critical argument integrated in the 'story' format. The usual sequence of sections is Introduction, Materials and Methods, Results and Discussion and Conclusions, with occasional variations. . . . These sections correspond to the sequence of the research." Among the "rules" for the Introduction Huth lists: "Tell the reader why the research was started."[36]

Huth thus assumes that a natural correspondence exists between the order of argument and the sequence of research that enables both to be expressed within the same structure of "format and content" (a correspondence he schematizes by presenting these orders in three parallel columns of a table). That this correspondence is not to be taken literally, however, is suggested by the fact that in the passage I have just quoted, "story" is twice placed in quotation marks. The same ambiguous message recurs in Huth's directions for the "Materials and Methods Section"

> How did you carry out your research? The critical reader will want to know exactly what you did and in enough detail to be able to judge whether the findings reported in the Results section are reliable support for your conclusions.[37]

If this instruction so far seems to suggest that an accurate historical narrative is called for, the paragraph finishes by partially withdrawing that injunction: "The Materials and Methods section should follow the sequence in which the research was planned even though not all research is executed exactly as planned." Under the heading of "Length," Huth advises authors about the problem of a paper's proportions:

> What do you do if the study was long and complex. . . ? Many of your readers may not want to read through all of the detail. One solution is to write a synoptic Materials and Methods section in which you give only the main points of design and procedure, with a more detailed description . . . as an appendix.[38]

As these directions strongly hint, the "story" that authors of modern research papers tell is not expected to be a narrative such as we try to tell as historians. It is a synopsis: a story reduced to the elements deemed essential to its outcome, pared not only of contingent circumstances encountered

---

36. Edward J. Huth, *How to Write and Publish Papers in the Medical Sciences* (Philadelphia: ISI Press, 1982), 47–49, 51.

37. Ibid., 53.

38. Ibid., 55.

along the way, but of all details of procedure and background that readers sharing the author's professional expertise will be able to supply from their own experience. Such transformations of investigative narratives into conventionalized stories more suitable to the purposes of the research paper are not only characteristic of present-day science: they originated with the research paper as a literary form.

7. Eighteenth-Century Medical Education and the Didactic Model of Experiment

IT WAS AN EIGHTEENTH-CENTURY COMMONPLACE that experiments were necessary to the progress of science, and one that was frequently heard at Edinburgh University. The university was a major center for medical education in the second half of the eighteenth century, and several of the medical professors, such as Alexander Monro *secundus*, Joseph Black, Daniel Rutherford, and John Leslie, were renowned for their scientific work throughout Great Britain and the Continent. Others, like William Cullen and Thomas Charles Hope, were less well known for their contributions to original research, but nonetheless inspired many students by their lectures and demonstrations.[1] The value attached to the creation of new knowledge by students was best expressed by Samuel Allvey in an essay written for the Royal Medical Society, the main student society. He wrote that in members' essays "it is encumbent on every member . . . to contribute those observations which he has made, and the opinions, which he has formed; rather than to collect the facts and opinions of others."[2]

Edinburgh students did not critically examine the concepts of "observation" and "experiment" so lauded by their contemporaries, and this is

1. Several recent works dealing with research at the University of Edinburgh are Lisa Rosner, "Students and Apprentices: Medical Education at Edinburgh University, 1760–1810." (Ph.D. diss., Johns Hopkins University, 1985); C. J. Lawrence, "Medicine as Culture: Edinburgh and the Scottish Enlightenment," (Ph.D. diss., University of London, 1984); A. L. Donovan, *Philosophical Chemistry in the Scottish Enlightenment* (Edinburgh: Edinburgh University Press, 1975). See also a series of articles by J. B. Morrell on the difficulties of establishing research traditions in the Scottish Universities: "The Chemical Breeders: The Research Schools of Liebig and Thomas Thomson," *Ambix* 19 (1972): 1–35; "The University of Edinburgh in the Late Eighteenth Century: Its Scientific Eminence and Academic Structure," *Isis* 62 (1970): 158–71; "Practical Chemistry in the University of Edinburgh," *Ambix* 16 (1969): 66–80.

2. Samuel Allvey, "Erysipelas," Royal Medical Society (RMS) Dissertations, 1786–1787, 20: 279. These unpublished dissertations are currently held in the library of the Royal Medical Society of Edinburgh, Student Centre, Bristo Square, Edinburgh. A microfilm copy of them is also available at the National Library of Medicine, Bethesda, Maryland.

hardly surprising. What is surprising is that for all their rhetoric, most students in the late eighteenth century did not even carry out their own investigations or contribute their own observations. There were a few prominent exceptions, such as Joseph Black and Daniel Rutherford, who both carried out important experiments for their MD theses.[3] Still, Edinburgh University, despite the eminence of its faculty, did not develop a research tradition similar to that of the Oxford physiologists a century earlier, or French chemists a generation later.[4]

The high value placed on experiments at Edinburgh, coupled with the small number of students who did them, has attracted the attention of historians. The most influential explanation was given by J. B. Morrell, who pointed to the lack of incentives for professors to offer classes in experimental technique. Faculty had to provide their own facilities for experiment, and apparatus was expensive; they could hardly have afforded to provide apparatus for an entire class. In addition, professors were paid by student fees, rather than a set salary. It was clearly in their own interest to teach large lecture classes, which would attract and could accommodate more students than a class in experimental technique.[5] Although this explanation accounts for why experimental classes were not incorporated into university lectures, this cannot be the whole story, for other kinds of small, experience-based classes were taught at Edinburgh, by either professors or outside lecturers. The university anatomy course did not provide students opportunities for dissection, but several outside lecturers gave practical anatomy classes which did.[6] University clinical lectures were frequently too crowded for students to see many patients, but extra-academical clinical lectures were offered in both medicine and surgery.[7] The lecturers in all these cases had to provide their own equipment for students, be it cadavers or patients. Why, then, did not even extra-academical lecturers offer experimental chemistry or physiology? There was certainly no shortage of what one student referred to as "hungry lecturers" looking for ways to attract students.[8] Had

3. Joseph Black, *De humore acido a cibis orto, et magnesia alba* (Edinburgh, 1754); Daniel Rutherford, *De aere fixo* (Edinburgh, 1772).

4. Robert G. Frank, *Harvey and the Oxford Physiologists: Scientific Ideas and Social Interaction* (Berkeley: University of California Press, 1980); Robert Fox, *The Caloric Theory of Gases: From Lavoisier to Regnault* (Oxford: Clarendon Press, 1971).

5. This is discussed in Morrell, "University of Edinburgh," and "Practical Chemistry."

6. J. D. Comrie, *History of Scottish Medicine*, 2 vols. (London: Wellcome Historical Medical Museum, 1932), 2: 628–631; Sylas Neville, *The Diary of Sylas Neville*, ed. Basil Cozens-Hardy (London: Oxford University Press, 1950), 205.

7. Rosner, "Students and Apprentices," 66, 245–249.

8. James Rush to Benjamin Rush, Edinburgh, 23 October 1809, Rush Papers Box 11, Yi 2, 7404 F21a, Library Company of Philadelphia Mss, Historical Society of Pennsylvania, Philadelphia.

there been student demand for courses in experiment, surely someone would have offered them.

Morrell's explanation embodies a sophisticated analysis of the impact of market factors on the teaching of science. It assumes, though, that both the concept of experiment and the methods of teaching it are unproblematic. This paper argues that it is precisely because that concept and those methods are so problematic that few examples of what we would consider student experimentation existed at Edinburgh University in the late eighteenth century. Experiment was defined and constrained by both the rhetoric with which it was described and the didactic models within which it was presented.

The first step in this analysis is to uncover what students thought of experiment. Students expressed their ideas most vividly in the essays on experimental subjects written for both the Royal Medical Society and the Chemical Society. The former, founded in 1737, was a medical version of the many literary and philosophical societies that flourished in Edinburgh, and the most important of the student societies.[9] According to one student, the "generous emulation to excel, [and] the desire for honorable fame"[10] were important motivations for joining the Society, and members considered themselves the intellectual elite of the student body. Each member was required to submit to the Society a clinical case history and a question on a medical or philosophical subject; once these were collected, they would be randomly drawn by the senior members, who would write an essay on each. Those essays were then presented to the Society as a whole and hotly debated. In the 1780s and 1790s, especially, the essays on medical or philosophical subjects often included accounts of experiments. The Chemical Society was of much shorter duration, apparently lasting only from 1785 to 1787. Its members too wrote essays, which resembled those of the Royal Medical Society in form, although limited, naturally, to chemical subjects. It appears to have been in part sponsored by Joseph Black.[11] We might

9. William Stroude, "History of the Royal Medical Society," in *List of Members, Laws, and Library-Catalogue, of the Medical Society of Edinburgh* (Edinburgh: William Aitken, 1820); James Gray, *History of the Royal Medical Society 1737–1937* (Edinburgh: Edinburgh University Press, 1952); see also D. D. McElroy, *Scotland's Age of Improvement: A Survey of 18th Century Literary Clubs and Societies* (Pullman: Washington State University Press, 1969); Rosner, 258–327.

10. Stroude, xii.

11. "Dissertations Read Before the Chemical Society, Instituted in the Beginning of the Year 1785," 1 vol., Chemical Library, University of Edinburgh (hereafter referred to as Chemical Society Dissertations). These essays are undated, though "AD 1787" is stamped on the front cover. Presumably the essays were written between 1785 and 1787 and bound in their present form in 1787. It is not clear whether the Society continued beyond that date. See also Gwen Averley, "The 'Social Chemists': English Chemical Societies in the Eighteenth and Early

expect, then, that his careful series of experiments on magnesia alba, or his theory of latent heat, would have inspired students in both Societies to carry out chemical experiments of their own.

In fact, even though many members of both Societies wrote essays on chemical topics, few carried out experiments. For example, James Home, Philip Elliott, and William Reid all discussed the existence of phlogiston in the 1780s and early 1790s, but although Elliot conceded that

> the numerous and valuable discoveries lately made, concerning the properties of different elastic fluids, and the great rage at present prevailing for experiments, have tended to elucidate [the phenomena] with great perspicuity,[12]

none of the three either attempted to replicate other experiments in the literature or developed new ones.[13] Other members discussed contemporary problems in chemistry, such as the composition and properties of fixed air, oxygen, and nitrogen. Each essay carefully described and analyzed experiments from Antoine-Laurent Lavoisier, Joseph Priestley, Nicholas Louis Vauquelin, and Humphry Davy, but none of these students attempted to reproduce them.[14]

Another student, John Latham, did give an account of an attempt to reproduce an experiment performed by Humphry Davy. In his essay, "What are the effects and mode of operation of . . . Nitrous Oxyd," Latham described the inhalation of nitrous oxide that he had witnessed, although not participated in, the preceding summer. He observed that two or three of the students "became quite purple in the face after breathing [the gas] a short time."[15] He also included in the essay the information he had re-

---

Nineteenth Century," *Ambix* 33 (1986): 99–100; J. Kendall, "The First Chemical Society, the First Chemical Journal, and the Chemical Revolution," *Proceedings of the Royal Society of Edinburgh* 63A (1952): 346–358, 385–400.

12. Philip Elliot, "How are the processes called phlogistic effected?" RMS Dissertations (1788–1789), 23: 280.

13. Philip Elliot (1788–1789), 23: 273–281; James Home, "What are the more general effects which phlogiston hath upon those bodies with which it is combined?" RMS Dissertations (1778), 11: 409–421; William Reid, "Whether or not is there in bodies, such a thing as the principle of fire?" (1793–1794), 31: 73–101.

14. John Begg, "What is the composition of fixed air?—its deleterious and salutiferous qualities?" RMS Dissertations, (1788–1789), 23:214–224; Francis Skrimshire, "What are the properties of oxygen and nitrogen and the substances resulting from their combination with each other?" (1798), 37:132–145; John W. Turner, "Is any nitrogen absorbed from the air in respiration?" (1783–1784), 15:273–299; Samuel Miller, "By what means is azote introduced into the animal body? And, how is it expelled from it?" (1802–1803), 48: 245–253; James Keir, "What are the chief properties of oxygen?" (1798–1799), 42:265–295; John William Stirk, "What are the properties of oxygen?" (1810–1811), 65:29–38.

15. John Latham, "What are the effects and mode of operation of the gazeous oxyd of azote, or nitrous oxyd?" RMS Dissertations (1801–1802), 46:413.

ceived from the students that, after inhaling, "the hearing and sight become more acute—Ideas seemed to pass with inconceivable rapidity thro' their minds—other indulged in an irresistible propensity to laugh."[16]

Latham's essay reveals a fundamental characteristic of student essays on experimental subjects: students were equally ready to refer in essays to experiments they had seen in lectures, to those they had read about in books, and to those they had performed themselves. J. B. Freer, for example, in an essay on Lavoisier's theory of caloric, devoted part of the essay to discussing whether white or black surfaces radiate more heat. In support of his own theory, which was that white surfaces radiate more heat, he mined the available chemical literature for experiments which would prove his point. Incidentally—but seemingly only incidentally—he also included a few of his own. There is certainly no evidence from his essay that he considered his own experiments more valuable as evidence than those of others.[17]

Further support for this observation comes from the fact that students almost never reproduced an experiment found in the literature. John Latham's account was an exception, but one suspects that the fact that nitrous oxide proved "a most agreeable source of new and highly delightful sensations (certainly, by all accounts, far superior to any ever produced by wine)"[18] was a factor in inducing students to perform Davy's experiment—there were no such attempts to reproduce Priestley's experiments on oxygen or nitrogen. No doubt similar factors explain why experiments on the properties of opium were discussed with so much enthusiasm.[19] Generally, if an experiment had been done once and reported by a reliable observer, students apparently felt that there was no need to do it again. Although they would all have agreed on the importance of experiments, what seemed to be comparatively unimportant was that they themselves perform them.

This attitude was not limited to chemical experiment. It can also be found in the records of the Experimental Committee, set up by the Royal Medical Society in 1785 to carry out physiological experiments, which were

16. Latham, 46:410–411.
17. J. B. Freer, "Does Lavoisierian theory of combustion account for the phenomena, or may they not be more satisfactorily explained?" RMS Dissertations (1805–1806), 54:279–284.
18. Latham, 46:420.
19. Edward Harrison, "What are the agreements and differences in the effects of opium and alcohol on the human body, in health and disease?" RMS Dissertations (1782–1783), 14: 188; Thomas Skeete, "What are the diseases in which opium may be employed with advantage, and how far can the modus operandi be accounted for?" (1783–1785), 17:110; William Alexander, "What are the effects of opium upon the healthy and diseased animal?" (1785–1786), 19:204. Opium was also a standard medication, as well as a subject of literary interest.

then presented to the Royal Medical Society as a whole. The Committee seems to have lasted for only about a year.

Some sense of the experimental technique can be gained from one series, entitled "Experiments with heat," which was begun on March 25, 1785. "At 20 Minutes past 2 o'clock," the records begin, "a very young white rabbit was put into water, heated to 80° of Farenh[ei]t. . . . At 25 minutes past 2, the vessel was plac'd on a pan of burning coals."[20] A record was kept of the time, the temperature within the vessel, and the reaction of the rabbit. At 2:36, the temperature was at 115°, and the "animal struggled a little"; by 2:37, temperature at 120°, it "squeaked violently." By 2:38 the temperature was at 122°, and the rabbit "gasped violently and seemed convulsed"; by 2:40 3/4, temperature 128°, respiration was very irregular; and by 2:42—the temperature remaining the same—the animal was "apparently dead and rigid." The records go on to say that the rabbit was removed and dissected, with special attention paid to the irritability of the heart. That, however, was almost the only comment made on the dissection; it apparently was less important to record than the rabbit's reaction to hot water.

Another series of experiments had to do with the effects of cold. These experiments basically involved freezing rabbits to death, either by putting them into water and then surrounding it with ice, or putting them into a mixture of ice and water. Again the rabbits were dissected, and again special attention was paid to the condition of the heart.

The rest of the experiments were more complicated. The next series, carried out on May 11, was designed "to discover the state of the stomach during vomiting."[21] A dog was given a purgative, and after it had taken effect his abdomen was cut open so that the stomach and diaphragm could be seen moving. One experiment failed because the dog inexplicably died instead of vomiting.[22]

It is hard to know how to interpret these experiments. The students nowhere give any indication of what they were intended to prove or disprove. That, in and of itself, is not unusual in accounts of eighteenth-century experiments. Even the notebooks of as careful an observer as Lavoisier can be hard to interpret, both because of the complexity of the phenomena he recorded and because his research aims were not always well defined at the outset.[23] Still, the Royal Medical Society experiments were

20. RMS, "Experimental Committee 1785," 17.
21. Ibid, 39.
22. Ibid.
23. Frederic Lawrence Holmes, *Lavoisier and the Chemistry of Life: An Exploration of Scientific Creativity* (Madison: University of Wisconsin Press, 1985), 13; F. L. Holmes, "Scien-

very unsophisticated in comparison with those of Marie-François Xavier Bichat at approximately the same age.[24] Students seldom kept track of exact measurements, nor did they describe in detail the purpose of the different operations. For the most part, the students seem only to have been interested during the period of the experiment itself, and to have relied for later information on the animals' condition on the keepers brought in to look after them.[25] In one experiment, on blood transfusions in calves, students paid careful attention to the heartbeat, respiration, and condition of eyes during the operation, but "as nothing remarkable occurred the animal was removed." Information concerning the aftereffects of the operation on the calf, such as onset of spasms, appear to have been conveyed to the Society by the "woman who attended [the calf]."[26]

Students did not even perform the actual manual work of the experiments. Instead, they were carried out by Andrew Fyfe, the demonstrator for the anatomy class.[27] Fyfe performed all of the incisions or ligatures done while the animal was alive, as well as the postmortem dissections. In the experiments to determine the effects of vomiting on the stomach, Fyfe not only made the incision, but was also the one to place his hand against the stomach and diaphragm to feel their motion.[28] This seems a very odd procedure for men interested in performing experiments: if the object was to do original research, why bring in someone else to do the actual work? Again, why did Edinburgh students place so little value on the aspect of experiment that we consider most important: actually carrying out the experiment itself?

The most probable answer is that the students were modeling their experiments on examples most familiar to them. One of these was the lecture demonstration. Students who attended the anatomy class were able to see the professor, Alexander Monro, explain some aspect of anatomy to his class, while Andrew Fyfe demonstrated the actual part of the body. This could well have been the model for the Society members, and one suspects that the experiments at the Royal Medical Society were done in lecture-demonstration style, one student explaining to the assembled group the object of the experiment, one student taking notes, and Fyfe performing

tific Writing and Scientific Discovery," *Isis* 78 (1987): 220–235.

24. John E. Lesch, *Science and Medicine in France: The Emergence of Experimental Physiology, 1790–1855* (Cambridge, Mass.: Harvard University Press, 1984), esp. 70–72.

25. RMS, "Experimental Committee 1785," 43.

26. Ibid.

27. Stroude, lxxxviii.

28. RMS, "Experimental Committee 1785," 39.

the actual operation. That would account for students' inattention to aftereffects of the operation: the lecture was over, so they saw no need to keep track of what happened to the animal. This was experiment, not as development of new knowledge, as confirmation of existing hypothesis, or even as practice for future investigation, but as practice for giving a lecture. Much that is puzzling in the records of the Experimental Committee becomes comprehensible if we do not take them to be modern—or even eighteenth-century—research reports, but rather use them to reconstruct what would have been a living model of experimental practice.

Students who attended Black's lectures would have seen him performing his own experiments. It is important to bear in mind, however, that his experiments were presented not as part of his ongoing research, but as demonstrations of chemical theory. That is, the Joseph Black most familiar to students was not the careful experimentalist of the early work on magnesia alba, but the learned professor, making use of experiments to explain some particular point of doctrine.[29] Remembering Black's lectures, one student said,

> In one department of his lecture he exceeded any I have ever known, the neatness and unvarying success with which all the manipulations of his experiments were performed. His correct eye and steady hand contributed to the one; his admirable precautions, foreseeing and providing for every emergency, secured the other. . . . The long table on which the different processes had been carried on was as clean at the end of the lecture as it had been before the apparatus was planted on it. Not a drop of liquid, not a grain of dust remained.[30]

Black's experiments, in other words, were presented to students as a polished performance, with no sign of the long process necessary to perfect them. Apparently nothing unforeseen or unexpected ever occurred, and since the experiments were presented with "unvarying success" there was no reason for students to think there might be some purpose in repeating them. As Joshua Parr put it, in his essay to the Chemical Society, Black had so fully illustrated the existence of latent heat "by the unquestionable test of experiment, that we need the aid of no other authority for it, nor ingenuity for its explanation." Parr went on to explain why his own account was taken

29. Henry Guerlac, "Joseph Black's Work on Heat," in *Joseph Black 1728–1799: A Commemorative Symposium,* ed. A. D. C. Simpson (Edinburgh: The Royal Scottish Museum, 1982), 14.

30. Henry, Lord Brougham, *Lives of Philosophers of the Time of George III* (London, 1855), 21, cited in R. G. W. Anderson, "Joseph Black: An Outline Biography," in Simpson, 11.

largely from Black's lectures: "I cannot propose anything new, nor can I conceive any illustration to be more perfect."[31] It might, then, not occur to Parr, or other students, to repeat similar kinds of experiments, so admirably described in the literature.

Indeed, the whole tenor of scientific literature, as it had developed over the previous century, would have reinforced this reliance upon written accounts of experiments. Peter Dear has shown the importance in scientific literature of locating an experiment in a specific time and place, and Steven Shapin and Simon Schaffer have analyzed the use of what they call "virtual witnessing," which "involves the production in a reader's mind of such an image of an experimental scene as obviates the necessity for either direct witness or replication."[32] Clearly one result of this tradition was that it did obviate the necessity for students to reproduce the experiments they read about. Indeed, it seems not to have occurred to them—or to many other eighteenth-century writers[33]—that the perception of the experimenter might have affected his account of the experiment itself, although they clearly recognized that it would have affected the conclusions he drew from it. In their own essays students made a clear distinction between "hypotheses" and "facts." The former were to be avoided at all costs, but the latter were clearly recognizable and straightforward. Thomas Burnside, for example, in his essay "On Zinc," claimed proudly, "The chemical philosophers of the present age are too enlightened, to be captivated with fanciful hypotheses, veiled in mystery, and clothed with the unintelligible terms of bombastic alchemy." What he himself wrote "is merely a collection of facts taken from the writings of the most celebrated chemists who have mentioned this subject."[34] This distinction was common in eighteenth-century writing, and it was encouraged by the essay genre in the Royal Medical Society because topics were submitted as questions, which required an author to state his opinion and support it with evidence in the face of sometimes tumultuous debate. Although we have no record of the debates on experimental essays, other accounts of student disputes suggest that the author's interpretation of experimental evidence—or his ability to interpret at all— were called into question more frequently than the evidence itself.[35]

31. Joshua Parr, "On Latent Heat," Chemical Society Dissertations, 149.

32. Peter Dear, "*Totius in Verba*: Rhetoric and Authority in the Early Royal Society," *Isis* 76 (1985):145–161; Steven Shapin and Simon Schaffer, *Leviathan and the Air-Pump: Hobbes, Boyle, and the Experimental Life* (Princeton, N.J.: Princeton University Press, 1985), 60.

33. Humphry Davy, for example, apparently saw no reason to repeat other people's experiments. *Dictionary of Scientific Biography*, s.v. "Davy, Humphry," by David M. Knight.

34. Thomas Burnside, "On Zinc," Chemical Society Dissertations, 357.

35. Debates in the Royal Medical Society, and among students generally, were frequently acrimonious. See Rosner, 305–313.

A third important influence on students' attitudes towards experiment was the main model of observational instruction available to them, the clinical case history. Case histories, written descriptions of diseases as they appear in the patient, were used at Edinburgh to test students' skills in diagnosis. A test of this kind was required for graduation, and, as mentioned earlier, members of the Royal Medical Society wrote essays diagnosing diseases from written case histories submitted by other members. The fundamental assumption behind this practice was that a well-written, accurate case history should be able to substitute for an actual bedside examination.

The rhetoric on the importance of bedside examination was much like the rhetoric on the importance of experiments. All students would have agreed on the necessity of actually seeing patients in order to gain practical medical experience. In the introduction to his clinical lectures, Professor John Gregory said,

> I have always been aware of the bad consequences of studying medicine from books and the lectures of professors. . . . A gentleman can declaim on the causes of diseases . . . but [if] he was never conversant with the sick, he is embarrassed with his erudition, and perplexed in cases, in which an apothecary's apprentice would find no difficulty.[36]

In practice, though, much of students' clinical experience came from studying written case histories. This is partly because university clinical lectures were extremely crowded,[37] but also because of the common assumption that notes from the clinical lectures, including Gregory's own, were of just as much value as the course itself. Sylas Neville, for example, recorded in his diary in 1774 that he had just made a "copy accurate and complete as the original" of a transcript of Gregory's clinical lectures from 1771, which he had borrowed from another student.[38] Neville could have attended another clinical course or arranged to accompany a surgeon in his practice to get more clinical experience; his decision to copy lecture notes instead reflects his belief that any careful eyewitness account could substitute for his own.

It was not only university teaching that blurred the distinction between

36. John Gregory, "Notes from Dr. Gregory's Clinical Lectures in the Royal Infirmary 1771," RMS Mss, 3–4. A note on the front cover incorrectly attributes the lectures to James Gregory.
37. David Skene to Andrew Skene, Edinburgh, 13 November 1751, Skene Papers, Aberdeen University Library, no. 40; Josef Frank, *Reise nach Paris, London, und Einem Grossen Theile des Uebrigen England und Schottlands in Bezeihung auf Spitaler, Versorgunghaeuser, und Gefaengnisse* (Vienna: Camesina, 1804), Pt. 2, 228.
38. Neville, 208.

bedside experience and written case histories. The common practice of medical consultation by mail rested on the assumption that a physician could base his diagnosis on a written account of symptoms as effectively as on a bedside examination. Indeed, a requirement for case histories presented to the Royal Medical Society was that they had to be "carefully narrated, as if wrote to a physician to be consulted." A further requirement was that each member could propose either "the case of any patient, which shall occur to him in practice," or a case from "the writing of any practical author, provided there be sufficient evidence that the case has really existed."[39] If we imagine those requirements transposed to essays on experimental subjects—that they had to be "carefully narrated, as if wrote to a philosopher to be consulted," and either consist of "any experiment, which shall occur to the author in practice," or else be drawn from "the writing of any practical author, provided there be sufficient evidence that the experiment has really existed"—we arrive at a very precise description of Edinburgh student experimentation.

The students' view of experiments, then—literally, their view of experiments—was of elegant demonstrations adduced to support Professor Black's theory of latent heat, of carefully framed accounts in the *Philosophical Transactions*, or of detailed case histories they themselves might receive from fellow physicians. In writing essays on experimental topics, they took their data from lecture demonstrations, from the experimental literature, or from their own experience, just as they took the data for clinical essays from clinical lectures, medical literature, or their own observations.

This view of experiment helps explain why Edinburgh University did not develop a sustained research tradition. The university could hardly be blamed for this, however; the proper way to teach experiment in the eighteenth century was no more obvious than it had been in the seventeenth. Experimental teaching, like experiment itself, was shaped by the available models for presentation of experiential knowledge. These models persisted well into the nineteenth century. As late as the 1860s, American students in Paris found it difficult to get practical research experience. Experiments even in Claude Bernard's or Charles-Edouard Brown-Sequard's classes were presented as parts of demonstrations, not ongoing investigations.[40]

39. "Regulations of a Society Instituted at Edinburgh for Improvement in Medical Knowledge 1737," bound with "List of Members of the Royal Medical Society of Edinburgh 1737–1811," RMS MSS. On consultation by mail, see Dorothy Porter and Roy Porter, *Patient's Progress: Doctors and Doctoring in Eighteenth-Century England* (Stanford: Stanford University Press, 1989), 76–78.

40. Robert G. Frank, "American Physiologists in German Laboratories, 1865–1914," in *Physiology in the American Context,* ed. Gerald L. Geison (Bethesda, Md.: American Physiological Society, 1987), 16–17.

This does not mean, of course, that students educated in this tradition, at Edinburgh or elsewhere, never did experiments. Edinburgh University did provide both a view of experiment as essential to the progress of science, and examples of how it might be done. At its best, it could, and did, inspire some students to go behind the scenes, perhaps to Black's laboratory, and find out for themselves how the demonstrations worked. One such student, Edmund Goodwyn, published his MD thesis on the physiology of drowning, which was highly enough regarded for Bichat to repeat Goodwyn's experiments in his own course.[41] Goodwyn began the book with a general account of his method. He had, he said,

> procured a large transparent glass bell, that would allow me to distinguish accurately the circumstances that took place within it; and when it was inverted, and filled with water, I put into it, at different times, several cats, dogs, rabbits, and other smaller animals, and confined them there 'till they had the appearance of being dead. As soon as they were put under the bell, I attended to the changes that took place in the body; and when they had lost the external signs of life, I opened the head, the breast, and belly, and examined the internal parts.[42]

Goodwyn was a member of the Royal Medical Society in 1785, and the beginning of his account is clearly similar to the records of the Experimental Committee. It seems therefore "not improbable," as Edinburgh students said about their theories, that Goodwyn's investigations were shaped by the didactic models of experiment at Edinburgh University.

The teaching of science has been particularly subject to the division between "social" and "intellectual" history of science. The former has concentrated on the political, economic, and institutional factors that affect teaching, whereas the latter has focused largely on professors' ideas, on the assumption that these are transmitted unambiguously through lectures to lecture notes. This paper has attempted to present an alternative by looking at the models within which ideas and information are presented. The visible practice of the lecture demonstration conveyed information as surely as the content of the lectures; the genre of the experimental essay carried with it assumptions about the neutrality of facts as well as the facts themselves; the

41. Lesch, 55.
42. Edmund Goodwyn, *The Connexion of Life with Respiration; or, an experimental inquiry into the effects of submersion, strangulation, and several kinds of noxious airs, on living animals: with an account of the nature of the disease they produce; its distinction from death itself; and the most effectual means of cure* (London: J. Johnson, 1788), section 1, "To ascertain the general effects of submersion on living animals."

form of the case history embodied a framework for writing about experiential knowledge in addition to the specific case history. What other lessons did the presentation of science teach? In the context of this book, that question is not merely rhetorical.

Selected Bibliography

THIS LISTING contains secondary material cited in the foregoing chapters bearing most directly on the collective themes of the volume. It is not meant to be exhaustive of the relevant literature.

Anderson, Wilda C. *Between the Library and the Laboratory: The Language of Chemistry in Eighteenth-Century France*. Baltimore and London: Johns Hopkins University Press, 1984.

Batens, Diderick, and Jean Paul Van Bendegem, eds. *Theory and Experiment: Recent Insights and New Perspectives*. Dordrecht: D. Reidel, 1988.

Bazerman, Charles. *Shaping Written Knowledge: The Genre and Activity of the Experimental Article in Science*. Madison: University of Wisconsin Press, 1988.

Bechler, Zev. "Newton's 1672 Optical Controversies: A Study in the Grammar of Scientific Dissent." In Yehuda Elkana, ed., *The Interaction Between Science and Philosophy*, 115–142. Atlantic Highlands, N.J.: Humanities Press, 1974.

Benjamin, Andrew E., Geoffrey N. Cantor, and John R. R. Christie, eds. *The Figural and the Literal: Problems of Language in the History of Science and Philosophy, 1630–1800*. Manchester: Manchester University Press, 1987.

Bitzer, Lloyd F. "The Rhetorical Situation." *Philosophy and Rhetoric* 1 (1968): 1–14.

Brannigan, Augustine. *The Social Basis of Scientific Discoveries*. Cambridge: Cambridge University Press, 1981.

Burke, Peter, and Roy Porter, eds. *The Social History of Language*. Cambridge: Cambridge University Press, 1987.

Bylebyl, Jerome. "Disputation and Description in the Renaissance Pulse Controversy." In A. Wear, R. K. French, and I. M. Lonie, eds., *The Medical Renaissance of the Sixteenth Century*, 223–245. Cambridge: Cambridge University Press, 1985.

Callon, Michel, John Law, and Arie Rip, eds. *Mapping the Dynamics of Science and Technology*. London: Macmillan, 1987.

Cantor, Geoffrey. "Light and Enlightenment: An Exploration of Mid-Eighteenth-Century Modes of Discourse." In David C. Lindberg and Geoffrey Cantor, *The Discourse of Light from the Middle Ages to the Enlightenment*, 67–106. Los Angeles: Clark Memorial Library, 1985.

———. "The Rhetoric of Experiment." In Gooding et al., *The Uses of Experiment*, 159–180.

―――. "Weighing Light: The Role of Metaphor in Eighteenth-Century Optical Discourse." In Benjamin et al., *The Figural and the Literal*, 124–146.

Chartier, Roger. *Cultural History: Between Practices and Representations*. Ithaca, N.Y.: Cornell University Press, 1988.

Christie, J. R. R., and J. V. Golinski. "The Spreading of the Word: New Directions in the Historiography of Chemistry 1600–1800." *History of Science* 20 (1982): 235–266.

Cohen, Ralph. "History and Genre." *New Literary History* 17 (1986): 203–219.

Coleman, Bob. "Science Writing: Too Good to Be True?" *New York Times* Book Review, September 27, 1987, 7.

Crane, Diana. "The Gatekeepers of Science: Some Factors Affecting the Selection of Articles for Scientific Journals." *American Sociologist* 2 (1967): 195–201.

Crosland, Maurice. *Historical Studies in the Language of Chemistry*. Cambridge, Mass.: Harvard University Press, 1962; New York: Dover, 1978.

Dagognet, François. *Tableaux et langages de la chimie*. Paris: Editions du Seuil, 1969.

Davis, Philip J., and Reuben Hersh. "Rhetoric and Mathematics." In Nelson et al., *The Rhetoric of the Human Sciences*, 53–68.

Dear, Peter. "Jesuit Mathematical Science and the Reconstitution of Experience in the Early Seventeenth Century." *Studies in History and Philosophy of Science* 18 (1987): 133–175.

―――. "*Totius in verba:* Rhetoric and Authority in the Early Royal Society." *Isis* 76 (1985): 145–161.

Dubrow, Heather. *Genre*. London and New York: Methuen, 1982.

Fleck, Ludwig. *Genesis and Development of a Scientific Fact*. Chicago: University of Chicago Press, 1979.

Foucault, Michel. *The Archaeology of Knowledge and the Discourse on Language*. Translated by A. M. Sheridan Smith. New York: Pantheon Books, 1972.

―――. *The Order of Things: An Archaeology of the Human Sciences*. New York: Pantheon Books, 1971.

Fowler, Alastair. *Kinds of Literature: An Introduction to the Theory of Genres and Modes*. Oxford: Clarendon Press, 1982.

Frye, Northrop. *Anatomy of Criticism*. Princeton: Princeton University Press, 1957.

Garfinkel, Harold. *Studies in Ethnomethodology*. Cambridge: Polity Press, 1984.

Golinski, Jan. "Language, Discourse and Science." In R. C. Olby, G. N. Cantor, J. R. R. Christie, and M. J. S. Hodge, eds., *Companion to the History of Modern Science*. London and New York: Routledge and Kegan Paul, 1990, 110–123.

―――. "Peter Shaw: Chemistry and Communication in Augustan England." *Ambix* 30 (1983): 19–29.

―――. "Robert Boyle: Scepticism and Authority in Seventeenth-Century Chemical Discourse." In Benjamin et al., *The Figural and the Literal*, 58–82.

Gooding, David, Trevor Pinch, and Simon Schaffer, eds. *The Uses of Experiment: Studies in the Natural Sciences*. Cambridge: Cambridge University Press, 1989.

Graham, Loren, Wolf Lepenies, and Peter Weingart, eds. *Functions and Uses of Disciplinary Histories*. Dordrecht: D. Reidel, 1983.

Grant, Edward. "Aristotelianism and the Longevity of the Medieval World View." *History of Science* 16 (1978): 93–106.

Hannaway, Owen. *The Chemists and the Word: The Didactic Origins of Chemistry.* Baltimore and London: Johns Hopkins University Press, 1975.

Harlan, David. "Intellectual History and the Return of Literature." *American Historical Review* 94 (1989): 581–609.

―――. "Reply to David Hollinger." *American Historical Review* 94 (1989): 622–626.

Hoffmann, Roald. "Die chemische Veröffentlichung—Entwicklung oder Erstarrung im Rituellen?" *Angewandte Chemie* 100 (1988): 1653–1663.

Hollinger, David A. "The Return of the Prodigal: The Persistence of Historical Knowing." *American Historical Review* 94 (1989): 610–621.

Holmes, Frederic L. *Lavoisier and the Chemistry of Life: An Exploration of Scientific Creativity.* Madison: University of Wisconsin Press, 1985.

―――. "Scientific Writing and Scientific Discovery." *Isis* 78 (1987): 220–235.

Horn, Ewald. "Die Disputationen und Promotionen an den deutschen Universitäten vornehmlich seit dem 16. Jahrhundert." *Beihefte zum Centralblatt für Bibliothekswesen* 4 (1893–1894): 1–126.

Hull, David. *Science as a Process: An Evolutionary Account of the Social and Conceptual Development of Science.* Chicago: University of Chicago Press, 1988.

Hunt, Lynn, ed. *The New Cultural History.* Berkeley: University of California Press, 1989.

Huth, Edward J. *How to Write and Publish Papers in the Medical Sciences.* Philadelphia: ISI Press, 1982.

Hyman, Stanley. *The Tangled Bank: Darwin, Marx, Frazer and Freud.* New York: Atheneum, 1962.

Jameson, Fredric. "Magical Narratives: On the Dialectical Use of Genre Criticism." In Jameson, *The Political Unconscious,* 103–150. Ithaca, N.Y.: Cornell University Press, 1981.

Jamieson, Kathleen M. "Generic Constraints and the Rhetorical Situation." *Philosophy and Rhetoric* 7 (1973): 162–170.

Jauss, Hans Robert. "Literary History as a Challenge to Literary Theory." *New Literary History* 2 (1970): 7–37.

Jordanova, Ludmilla J., ed. *Languages of Nature: Critical Essays on Science and Literature.* London: Free Association Books, 1986.

Kaufmann, G. "Zur Geschichte der academischen Grade und Disputationen." *Centralblatt für Bibliothekswesen* 11 (1894): 201–225.

Kendall, J. "The First Chemical Society, the First Chemical Journal, and the Chemical Revolution." *Proceedings of the Royal Society of Edinburgh* 63A (1952): 346–358, 385–400.

Kitcher, Philip. *The Nature of Mathematical Knowledge.* Oxford: Oxford University Press, 1984.

Kronick, David. *A History of Scientific and Technical Periodicals: The Origins and Development of the Scientific and Technical Press 1665–1790.* 2d ed. Metuchen, N.J.: Scarecrow Press, 1976.

Kuhn, Thomas S. *The Structure of Scientific Revolutions.* 2d ed. Chicago: University of Chicago Press, 1970.

Kundert, Werner. *Katalog der Helmstedter juristischen Disputationen Programme und*

Reden 1574–1810. Repertorien zur Erforschung der frühen Neuzeit, Bd. 8. Wiesbaden: Otto Harrassowitz, 1984.

LaCapra, Dominick, and Steven Kaplan, eds. *Modern European Intellectual History: Reappraisals and New Perspectives.* Ithaca, N.Y.: Cornell University Press, 1982.

———. *Soundings in Critical Theory.* Ithaca, N.Y.: Cornell University Press, 1989.

Latour, Bruno. *Science in Action: How to Follow Scientists and Engineers Through Society.* Cambridge, Mass.: Harvard University Press, 1987.

Latour, Bruno, and Steve Woolgar. *Laboratory Life: The [Social] Construction of Scientific Facts.* Beverly Hills, Calif. and London: Sage, 1979; 2d ed. Princeton, N.J.: Princeton University Press, 1986.

MacLeod, Roy, and Paul Gary Werskey. "Is It Safe to Look Back?" *Nature* 224 (1969): 417–476.

Merton, Robert K. *The Sociology of Science: Theoretical and Empirical Investigations.* Edited by Norman J. Storer. Chicago: University of Chicago Press, 1973.

Miller, Carolyn R. "Genre as Social Action." *Quarterly Journal of Speech* 70 (1984): 151–167.

Miner, Earl. "Some Issues of Literary 'Species, or Distinct Kinds'." In Barbara K. Lewalski, ed., *Renaissance Genres: Essays on Theory, History, and Interpretation,* 15–44. Cambridge, Mass.: Harvard University Press, 1986.

Mulkay, Michael. *The Word and the World: Explorations in the Form of Sociological Analysis.* London: George Allen and Unwin, 1985.

Mulkay, Michael, and Nigel Gilbert. *Opening Pandora's Box: A Sociological Analysis of Scientists' Discourse.* Cambridge: Cambridge University Press, 1984.

Myers, Greg. *Writing Biology: Texts in the Social Construction of Scientific Knowledge.* Madison: University of Wisconsin Press, 1990.

Naylor, R. H. "Galileo's Experimental Discourse." In Gooding, *The Uses of Experiment,* 117–134.

Nelson, John S., Allan Megill, and Donald N. McCloskey, eds. *The Rhetoric of the Human Sciences: Language and Argument in Scholarship and Public Affairs.* Madison: University of Wisconsin Press, 1987.

Peterfreund, Stuart, ed. *Literature and Science: Theory and Practice.* Boston: Northeastern University Press, 1990.

Phillips, J. P. "Liebig and Kolbe, Critical Editors." *Chymia* 11 (1966): 89–98.

Pinch, Trevor J. "Toward an Analysis of Scientific Observation: The Externality and Evidential Significance of Observational Reports in Physics." *Social Studies of Science* 15 (1985): 3–36.

Rousseau, George S., ed. *Science and the Imagination.* New York: Annals of Scholarship, 1986.

Rudwick, Martin J. S. "Charles Darwin in London: The Integration of Public and Private Science." *Isis* 73 (1982): 186–206.

———. "The Emergence of a Visual Language for Geological Science, 1760–1840." *History of Science* 14 (1976): 149–195.

———. *The Great Devonian Controversy: The Shaping of Scientific Knowledge Among Gentlemanly Specialists.* Chicago: University of Chicago Press, 1985.

Sargent, Rose-Mary. "Scientific Experiment and Legal Expertise: The Way of Experience in Seventeenth-Century England." *Studies in History and Philosophy of Science* 20 (1989): 19–45.

Schatzberg, Walter, Ronald A. Waite, and Jonathan K. Johnson, eds. *The Relations of Literature and Science: An Annotated Bibliography of Scholarship, 1880–1980*. New York: Modern Language Association, 1987.

Schmitt, Charles B. "Experience and Experiment: A Comparison of Zabarella's View with Galileo's in *De motu*." *Studies in the Renaissance* 16 (1969): 80–138.

———. "Galileo and the Seventeenth-Century Text-Book Tradition." In Paolo Galluzzi, ed. *Novità celesti e crisi del sapere: Atti del Convegno Internazionale di Studi Galileiani*, 217–228. Firenze: Giunti Barbèra, 1984.

Schuster, John A. "Methodologies as Mythic Structures: A Preface to the Future Historiography of Method." *Metascience: Annual Review of the Australasian Association for the History, Philosophy and Social Studies of Science* 1/2 (1984): 15–36.

Schuster, John A., and Richard R. Yeo, eds. *The Politics and Rhetoric of Scientific Method: Historical Studies*. Dordrecht: D. Reidel, 1986.

Shapin, Steven. "Pump and Circumstance: Robert Boyle's Literary Technology." *Social Studies of Science* 14 (1984): 481–519.

Shapin, Steven, and Simon Schaffer. *Leviathan and the Air-Pump: Hobbes, Boyle and the Experimental Life*. Princeton, N.J.: Princeton University Press, 1985.

Sheets-Pyenson, Susan. "From the North to Red Lion Court: The Creation and Early Years of the *Annals of Natural History*." *Archives of Natural History* 10 (1981): 221–249.

"Symposium: Rhetoricians on the Rhetoric of Science." *Science, Technology, & Human Values* 14 (1989): 3–49.

Vickers, Brian. "The Royal Society and English Prose Style: A Reassessment." In Vickers and Nancy S. Struever, *Rhetoric and the Pursuit of Truth: Language Change in the Seventeenth and Eighteenth Centuries*, 3–76. Los Angeles: Clark Memorial Library, 1985.

Wear, Andrew. "William Harvey and the 'Way of the Anatomists'." *History of Science* 21 (1983): 223–249.

Whitley, Richard D. "The Operation of Science Journals: Two Case Studies in British Social Science." *Sociological Review* n.s. 18 (1970): 241–258.

Woolgar, Steve. "Discovery: Logic and Sequence in a Scientific Text." In Karin D. Knorr, Roger Krohn, and Richard Whitley, eds. *The Social Process of Scientific Investigation*, 239–268. Dordrecht: D. Reidel, 1981.

———. "What Is the Analysis of Scientific Rhetoric For? A Comment on the Possible Convergence Between Rhetorical Analysis and Social Studies of Science." *Science, Technology, & Human Values* 14 (1989): 47–49.

———. ed. *Knowledge and Reflexivity: New Frontiers in the Sociology of Knowledge*. London: Sage, 1988.

Contributors

Peter Dear is Assistant Professor of History at Cornell University. He has published articles on seventeenth-century science and philosophy in *Isis, Studies in History and Philosophy of Science, Journal of the History of Philosophy,* and elsewhere, and is a contributor to the forthcoming *Cambridge History of Seventeenth-Century Philosophy.* He is the author of *Mersenne and the Learning of the Schools* (Cornell University Press, 1988).

Thomas H. Broman completed his Ph.D. in History of Science at Princeton University in 1987. Currently an Assistant Professor in the Department of the History of Science, University of Wisconsin at Madison, he has published on German universities and medical education in the eighteenth century and is now studying the history of German physiological writing in the eighteenth and early nineteenth centuries.

Frederic L. Holmes is Chairman of the Section of the History of Medicine at Yale University and a former president of the History of Science Society. He is the author of a number of books, including *Claude Bernard and Animal Chemistry* (1974) and *Lavoisier and the Chemistry of Life* (1985), as well as many articles.

Bruce J. Hunt is Assistant Professor of History at the University of Texas, Austin. He has published articles on the history of nineteenth-century physics in *Isis, Historical Studies in the Physical Sciences, The British Journal for the History of Science,* and elsewhere, and is currently completing a revised version of his dissertation, "The Maxwellians," for publication by Cornell University Press.

Lynn K. Nyhart received her Ph.D. in the history and sociology of science from the University of Pennsylvania in 1986 and is now an Assistant Professor in the Department of the History of Science, University of Wisconsin at Madison. She specializes in the history of German biology in the nineteenth century, on which she is preparing a book.

Lissa Roberts received her Ph.D. in European Intellectual History at the University of California, Los Angeles, in 1985. She is now Assistant Professor of History at San Diego State University and is working on a book entitled *The Instrumental Space: A History of Eighteenth-Century Pneumatic Chemistry.*

Lisa Rosner received her Ph.D. from the Department of the History of Science at the Johns Hopkins University in 1985. She is currently Assistant Professor of History at Stockton State College, Pomona, New Jersey, and is working on a study of medical students at Edinburgh University, 1750–1824.

Index

This index provides references to all bibliographical citations.

This book has been set in Linotron Galliard. Galliard
was designed for Merganthaler in 1978 by Matthew
Carter. Galliard retains many of the features of a sixteenth
century typeface cut by Robert Granjon but has some
modifications which give it a more contemporary look.

Printed on acid-free paper.